# Criminal Justice
## of Decency

# Criminal Justice and the Pursuit of Decency

ANDREW RUTHERFORD

Foreword by
Lord Scarman

Oxford   New York

OXFORD UNIVERSITY PRESS

1993

Oxford University Press, Walton Street, Oxford OX2 6DP
Oxford New York Toronto
Delhi Bombay Calcutta Madras Karachi
Kuala Lumpur Singapore Hong Kong Tokyo
Nairobi Dar es Salaam Cape Town
Melbourne Auckland Madrid
and associated companies in
Berlin Ibadan

Oxford is a trade mark of Oxford University Press

British Library Cataloguing in Publication Data
Data available

Library of Congress Cataloging in Publication Data
Rutherford, Andrew, 1940–
Criminal justice and the pursuit of decency / Andrew Rutherford.
p.    cm.
Includes bibliographical references and index.
1. Criminal justice, Administration of—Great Britain—Moral and
ethical aspects.   2. Criminal justice personnel—Great Britain—
Interviews.   I. Title.
HV9960.G7R87  1993   364.941—dc20     92-1686
ISBN 0-19-215896-1
ISBN 0-19-285275-2 (pb)

1 3 5 7 9 10 8 6 4 2

Typeset by Best-set Typesetter Ltd.
Printed in Great Britain by
Biddles Ltd.
Guildford and King's Lynn

*For Judith*

# FOREWORD

by Lord Scarman

'Criminal Justice and the Pursuit of Decency'—an arresting title indeed. By choosing it Andrew Rutherford has given notice that in this work he is not, save indirectly, concerned with law, ethics, or the theory of punishment, but with 'decency'. One aspect of 'decency' is, of course, good manners in handling those put into your charge: perhaps he might have used as a secondary title to his book some words attributed to a fourteenth-century Lord Chancellor of England: 'Manners makyth man.'

Rutherford's message is simple and clear. Humane values must lie at the heart of the system. He puts the point succinctly in the Preface: 'the expression of humane values within criminal justice ultimately resides with practitioners. Indeed, in the absence of such values, criminal justice invariably descends into apathy, and ultimately, violence.' At this point, a warning to lawyers who, as I devoutly hope, will be reading this work. Rutherford uses the word 'practitioners' to describe all who work in the system whether they be legally qualified or not: the word includes police, probation officers, members of the Crown Prosecution Service, magistrates' clerks, court officials, and prison staff at all levels. He does not, of course, exclude judges and professional lawyers practising in the courts: but his purpose is to emphasize the importance to a civilized system of criminal justice of humane values at *all levels* of its administration.

Rutherford is, of course, a scholar of renown and the author of important works in the field of criminal justice. He is a man of wide learning: but it is of great significance that he is also an experienced practitioner in the field. He was a prison officer for some ten years. It is his combination of learning and experience that gives this, his latest published work, its importance. His views are based on a deep understanding of the subject. But the case in which he believes and which he asks us to accept does not depend exclusively on his scholarship and experience. In preparing this book he went to twenty-eight practitioners within our system of criminal justice whom he believed to share at least some of his views and invited them to talk to him. Their narratives are the core of the book, and supply much of the evidence upon which he relies in reaching his conclusions. Their narratives and his comments lend the book a persuasiveness which no amount of sociological analysis and argument could provide. The twenty-eight practitioners and Rutherford expose the urgent need for the pursuit of decency in our system of criminal justice.

Basing himself on his research and experience, he finds that the values and beliefs that shape the daily work and concerns of criminal justice practitioners fall into three clusters. These clusters become what he calls the 'working credos' of practitioners. They represent, in whole or in part, the practitioners' view of the purpose of their work. They are:

1. the punishment credo, which he describes as 'the punitive degradation of offenders'
2. the efficiency credo, which he describes as concentration upon issues of 'management, pragmatism, efficiency and expedience' and
3. the caring credo, which is an attitude towards suspects, accused persons, and prisoners based on 'liberal and humanitarian values'.

If, as Rutherford asserts with the support of substantial evidence, punitive degradation of the convicted is the essence of the punishment credo, the sooner it is replaced by humanitarian values the better. For such a credo carries with it echoes of Dante's Inferno—'the sorrowful city' where all who enter must abandon hope: that is carrying punishment to the extreme of unacceptable cruelty.

But, even if the punishment credo may well not go so far as 'punitive degradation' save in the minds of a minority of practitioners, Rutherford's argument that the punishment and efficiency credos, if unaccompanied by a caring attitude towards the individuals, be they suspects or convicts, in the charge of the system, will lead to 'apathy and ultimately violence' is founded on evidence which I find compelling.

Nevertheless some will suggest that Rutherford's fears are exaggerated. Is he alarmist? Or is he targeting a real danger? I have no doubt that he is targeting a real danger.

The evidence of the narratives of the practitioners whom he interviewed support his view that it is necessary for the morale of penal institutions to develop a mutual understanding between prison staff and inmates. And his research, as the book shows, bears this out.

There is, I believe, no need to be pessimistic, though we must remain vigilant.

So far as the work of the police, the prosecution, and the courts are concerned, there has been real progress since 1984 when the Police and Criminal Evidence Act was passed. And in 1986 the Crown Prosecution Service was established: this is perhaps the most hopeful innovation of all, ensuring as it does an independent professional control of all aspects of prosecution. The major disaster in this area has been the spate of cases of miscarriage of justice. But I am confident of progress here. The Runciman Commission's report is expected in 1993. And the Crown Prosecution Service is already developing an effective control of prosecution in serious cases. We may be sure of effective reforms in the system within a few years.

But the prisons are another matter. They remain Dante's 'sorrowful city'. Andrew Rutherford's pursuit of decency can, and should, encourage substantial advance in the development of a mutual basis of understanding between practitioners and inmates. I commend his book not only to those who labour to improve our system of criminal justice but also to the general public. Nobody should pass by on the other side.

SCARMAN

*July* 1992

# PREFACE

REMARKABLY little is known about the beliefs and sentiments that impact upon the work of criminal justice practitioners. That this remains relatively unchartered territory is surprising, given the profound social and political implications of the activities of personnel working at every stage of the criminal justice process.

My interest in this topic began in the early 1960s, when I joined the English prison service and was confronted by a broad diversity of values and aspirations among my colleagues on the assistant governors' staff course. Over the next ten years I was puzzled by the variations in the regimes in the three penal institutions within which I worked. The climate of the times, priorities of the system, and institutional traditions were clearly all considerations, but not necessarily more so, I felt, than the attitudes and objectives of the staff themselves.

After leaving the prison service in the early 1970s, I was able to pursue these themes at a rather safer distance. Teaching a course on 'Innovation and Criminal Justice' at the University of Minnesota in 1975, I invited several leading practitioners, from police chiefs to prison wardens, to share their professional experiences with the class. The success of this involvement of practitioners in the academic setting encouraged me to try out new opportunities as they arose in a variety of settings. Over recent years this interest has been further enhanced by an appreciation that the line between policy and practice has been too finely drawn. In particular, I have become increasingly impressed by the extent to which practitioners shape and lead criminal justice policy.

The term 'humane values' fails to capture precisely the working ideology that is the topic of this study. 'Benevolence', 'compassion', 'sympathy', and 'care' do not suffice and may in fact distract. 'Liberal humanitarianism', perhaps closest, but hardly a term that lends itself to easy usage, encompasses the cluster of values that manifest themselves in empathy and respect for the offender, an optimism that people can make progress with their lives, and an insistence upon justice and on clear lines of account-ability to democratic institutions.

It has become conventional to mistrust good intentions in criminal justice practice. Such is the catalogue of harm, pain, and, indeed, disaster resulting from benevolence that humane endeavours often are dismissed as a smoke-screen that conceals a different reality. Steven Marcus has put the issue in this way:

How is it that good people—decent upright and well meaning citizens—can contrive, when they act on behalf of others and in the name of some higher principle or of some benign interest, to behave so harshly, coercively and callously, so at odds with what they understand to be their good intentions?[1]

While the essence of this sentiment has some validity, the danger is that healthy wariness of humane motives slides into cynical and indiscriminate rubbishing. The underlying premise of this book is that the expression of humane values within criminal justice ultimately resides with practitioners. Indeed, in the absence of persons adhering to such values, criminal justice invariably descends into apathy and, ultimately, violence.

The twenty-eight persons whose narratives provide the sub-stance of this book are a remarkable group of men and women. The extracts from their narratives speak to their courage, forti-tude, and good humour. Exposing personal attitudes and beliefs is never without risk, and the ready willingness to be involved in the project typifies their approach to each day's work. While each person agreed to the publication of the passages from his or her narrative, the responsibility for their presentation here rests entirely with the author.

The transcribing of the taped interviews was diligently carried out by Damaris Inie. For additional typing, I must also thank Shani Turner. The major burden fell to Lucille Foster, who with conscientious good cheer saw the book through its many drafts. The dedication reflects the loving encouragement of Judith Rutherford.

A. R.

*Southampton*
*January 1992*

# CONTENTS

# 1 Working Credos

THE enterprise of criminal justice revolves around a host of competing purposes, beliefs, and values. It is shaped by the society of which it is a part, but society itself is also influenced by the values and practices of criminal justice.[1] To a large degree, this dynamic process determines what happens to persons suspected or convicted of criminal offences. These outcomes are crucial for the individuals concerned, but they also constitute a moral barometer for the wider society, serving to 'mark and measure the stored-up strength of a nation'.[2] But how do particular sets of values find expression in the work of criminal justice agencies?

Prompting this study's approach was a concern that humane values have a precarious existence in the criminal justice process, even within societies that have a liberal social and political tradition. When Jerry Miller says 'our side in this always loses', it is less a note of despair than a declaration of how essential it is that some people keep trying 'to make an inhuman system very decent'.[3] A sombre portrait of criminal justice emerges from a stream of reports of miscarriages of justice, prison scandals, and abuses of power in countless criminal justice agencies across the world.

One example encompasses many of these common themes and illustrates the two-way influence of values and beliefs between criminal justice agencies and the wider society. In April 1991, Hal Wootten QC, a member of the Royal Commission on Deaths of Australian Aboriginals in Police Custody, reported on his investigation of eighteen deaths in three states. Referring to one of these, a man in a state of alcohol withdrawal who had been placed in police custody, Mr Wootten's strictures encompassed both police officers and medical personnel: 'I find it impossible to believe that so many experienced people could have been so reckless in the case of a seriously ill person dependent on them, were it not for the dehumanised stereotype of Aboriginals...in that stereotype, a police cell is a natural and proper place for an Aboriginal.'[4] The pervasive but often concealed relationship between values and criminal justice has been cogently stated by Walter Miller: 'Ideology and its consequences exert a powerful influence on the policies and procedures of those who conduct the enterprise of criminal justice, and ... the degree and kinds of influences go largely unrecognised. Ideology is the permanent hidden agenda of criminal justice.'[5]

These ideological influences are exerted in three overlapping and interactive arenas. Firstly, there are the formal statements of policy which are the products of the legislature, the higher reaches of the civil service, and the judiciary in their law-making role. For legislators and judges, in particular, opportunities frequently arise to articulate the value choices being made. In contrast, these choices are rather more obscure with respect to the second arena, the structural arrangements of criminal justice. Formal procedures and other aspects of the organization tend to be shaped without explicit reference to underlying ideological preferences. This is especially likely with reference to the influences and effects of informal arrangements, subcultures, and traditional practices. Although these informal structural patterns have attracted considerable scholarly exploration, much of this research has contributed to an unduly deterministic view of criminal justice activity. Indeed, some of these studies appear to imply that the conduct of criminal justice is all but devoid of human actors.

This book is concerned mostly with the third arena, namely one in which the ideas and beliefs that are held by criminal justice personnel impact upon practice. Although research has only barely scratched the surface, it is clear that the relationship between ideology and practice is both complex and unpredictable.[6] That what people say they believe is by no means invariably matched by what they do was highlighted in a study of juvenile courts in Massachusetts in the early 1960s. Stanton Wheeler and colleagues reported that:

The judges who have taken the more severe actions are those who read more about delinquents, who read from professional journals, who do not wear their robes in court, and who are more permissive in outlook ... Severity of the sanctions, therefore, appears to be positively related to the degree to which the judge uses a professional, humanistic, social welfare ideology in making his decisions.[7]

Although these findings were 'weak and provisional', the researchers suggested that the judges may have regarded placement in institutions as 'benign, humane and therapeutic, rather than as existing as a last resort for punishment and community protection'.[8] That attachment to rehabilitative purposes may have been a clue to this paradox was supported by an association between the willingness of the judges to commit youngsters to institutions and their sensitivity to psychological disorder rather than to perceived seriousness of the acts for the community.

In a more extensive study of sentencing practice in Ontario, John Hogarth found a large number of statistically significant correlations between 'cognitive-complexity' and sentencing choices:

Magistrates who use suspended sentence frequently in indictable cases, particularly suspended sentence with probation, tend to discriminate better in complicated fact situations, are capable of bringing a larger body of information to bear in a given problem, and expend more effort in problem solving. In contrast, magistrates who rely heavily on fines do not appear to be as subtle in discriminating among information.[9]

Hogarth's main conclusion was that the frequent use of both very short and very long custodial sentences was associated with a simpler and more rigid way of dealing with information. But this finding was qualified by the willingness of 'cognitively complex' magistrates to sentence persons convicted of minor offences to reformatory institutions. The explanation, as in Wheeler's study, appeared to be the belief that offenders would actually benefit from a period of custody.

The values and beliefs that shape the daily work and professional careers of criminal justice practitioners fall into three clusters. The first of these embraces the punitive degradation of offenders. The second cluster speaks less to moral purpose than to issues of management; pragmatism, efficiency, and expediency are the themes that set the tone. Third, and the focus of this book, there is the cluster of liberal and humanitarian values. The delineation of these clusters, which for convenience are referred to as Credos One, Two, and Three, derives from a diverse body of research which dates mostly from the 1960s. As Francis Allen has observed, it was not until that decade that scholars seemed fully to recognize the 'inherently and inescapably political' basis of criminal justice; and, freed from the purposes and assumptions of the system, they were able to broaden the scope of their research activity.[10] For example, after scrutinizing the mores of a

California police department, Jerome Skolnick reported that officers thought the rules of criminal procedure to be irrational, viewing them with 'the administrative bias of the craftsman, a prejudice contradictory to the due process of law'.[11] The 'working personality' of the police officer, he argued, was more likely to be shaped by the prevailing ethos of the police organization than by the requirements of formal law. Skolnick's contacts with police officers revealed that 'a Goldwater type of conservatism was the dominant political and emotional persuasion of the police'. He encountered only three officers who described themselves as being politically liberal, and these persons insisted that they were exceptional.[12]

A generation of academic researchers immersed themselves in the daily activities of police officers, court officials, prosecutors, and prison staff. These studies, which highlighted the variety of formal and informal agendas that characterize criminal justice, were prompted by and in turn enriched new conceptual frameworks that took account of competing sets of values. In the pivotal work of this kind, the legal scholar, Herbert Packer, in seeking to give operational content to the underlying complex of values in criminal justice, identified two ideal types which he called the models of Crime Control and Due Process.[13] Packer's formulation should be set within the historical context of an activist phase of the Supreme Court, when, under Chief Justice Earl Warren, it sought to extend constitutional protections for individuals caught up in the criminal net. In particular, the rights of suspects with respect to arrest and interrogation were strengthened, and unlawfully obtained evidence had to be excluded by the courts. At this time, federal courts began to abandon their 'hands off' stance to the management of prisons;[14] the intervention of courts at various stages of the criminal process was applauded by liberals but enraged most conservative commentators and legislators and a great many law enforcement officials. This ideological divide forms the basis of Packer's contrast of the theoretical and operational features of the two models. Driving the Crime Control model is the repression of criminal conduct, with a high premium placed upon efficiency and on getting results. Inherent in Crime Control is the 'presumption of guilt', which means a prediction of outcome arising from the collective attitudes of practitioners. By contrast, in stressing the possibility of error, Due Process places reliability over efficiency. The doctrine of legal guilt and other intrinsic safeguards arising from the associated presumption of innocence are distinctive features, along with notions of equality and fairness. The model also implies a healthy scepticism as to the morality and utility of criminal punishment. If Crime Control suggests an assembly line, Due Process may be likened to an obstacle course. Although Packer asserted that it is the values of Due Process that are most deeply impressed on the criminal law, he concluded that for much of the time actual practice probably approximates more closely to the dictates of Crime Control.

The characterization of criminal justice, as a struggle between the formal requirements of Due Process and the more informal pursuit of Crime Control, has not gone without challenge. Notably, Doreen McBarnett, on the basis of her study of practice in English courts, argued that, far from embodying notions of legality, the law on criminal procedure institutionalizes deviation from that ideal. McBarnett concluded:

Judges and politicians may deal in the rhetoric of civil rights and due process, but the actual rules they create for law enforcement and the policies they adopt on sanctioning police malpractice are less about civil rights than about smoothing the path to conviction, less about due process than post-hoc acceptance of police activities as justifying themselves.[15]

Her insistence that both the law of books and the law in action reflect Crime Control values led her to castigate Packer's thesis as 'a false dichotomy'. McBarnett, however, overstates her objections. While there is much in statutory and case law that can be used to support her argument, there is also a great deal that works against it. Furthermore, she makes no acknowledgement of the dynamic and crucial role provided by practitioners who adhere to humane values.

That the structure of criminal justice is geared to 'smoothing the way to conviction' is reiterated in other studies of English courts. For example, Michael McConville and John Baldwin, impressed more by informal patterns than by the formal expectations of the law, concluded that the process is structured 'to produce convictions within a framework of uncertain protections, confused and ill-defined responsibility, in which informal and even unlawful behaviour of various participants is directly or indirectly legitimized'.[16] Furthermore, the arrangements seem 'to be structured in part to give rights which are unenforceable and in part to give rights which are in some measure enforceable but not enforced'. These structural flaws include 'selective investigation, inaccurate recording of evidence, prosecutions instituted for non-evidentiary reasons, imperfect screening of weak cases, indirect and ambiguous allocation of responsibilities between different groups, pressures placed upon defendants and lawyers, and the denial or emasculation of rights. These are not accidental features of an otherwise sound system, a system framed to avoid such shortcomings: they constitute the system itself.'[17] Finally, in another study, A. E. Bottoms and J. D. McClean encountered little evidence of Due Process values or, indeed, much sign of any struggle between competing value systems. Instead, they were impressed by a pervasive 'liberal bureaucratic' model which, despite superficial similarities with Due Process, was in practice much closer to the aims of Crime Control, and with which it had formed 'an effective alliance'.[18]

These studies of the criminal courts have contributed to a view of criminal justice practice that is characterized by collusive cohesion rather than tension and conflict. The informal agenda and structure are about getting through the day's business as expeditiously as possible rather than pursuing any moral mission. Robert Reiner has struck a similar note, with respect to policing, in observing that the police culture reflects 'a very pragmatic, concrete, down-to-earth, anti-theoretical perspective . . . a kind of conceptual conservatism'.[19]

The bureaucratization of much criminal justice activity during the twentieth century is significant in explaining how this pragmatic stance has come about. David Garland recently noted that the emergence of professional bureaucracies coincided with a period 'in which punitive sentiments have been increasingly marginalised in official discourses and replaced by more utilitarian objectives and expectations'.[20] Despite differences with respect to training, technical capacity, and matters of penal policy, he argued that penal practitioners see themselves as performing a positive and useful social task and 'tend to orient themselves towards institutionally defined managerial goals rather than socially derived punitive ones'.[21] For Garland, the contemporary ideological divide is between persons adhering to the objectives of smooth management and those seeking the social condemnation of offenders. He alluded to but did not explicitly allow for a third dimension: namely, the influences upon criminal justice of humane values.

## Method

I interviewed twenty-eight Practitioners holding senior positions across the criminal justice process between November 1988 and February 1991.[22] All but two of these persons were working in England and Wales, with one each holding a position in Scotland and Northern Ireland. Six were from the prison system and the Crown Prosecution Service and five from the police and probation services. The remainder consisted of two justices' clerks, a senior official in the Home Office, a solicitor, a leading figure with the Magistrates' Association, and a recorder. No full-time judges were included as it was anticipated that permission for them to take part would not be forthcoming.[23] Everyone approached agreed to be interviewed, although one person withdrew prior to the interview when it emerged that authority would need to be obtained from ministers. The interviewees are not intended to be representative of senior criminal justice practitioners. They were chosen because they were well known to me or to others I consulted as adhering in their work to liberal and humanitarian values. Their selection was not difficult, as the same names tended to be mentioned whenever enquiries were made. But it was also readily apparent that the persons interviewed represent a rather small and distinct minority of the British criminal justice élite.[24]

All the Practitioners had worked within criminal justice agencies for at least fifteen years, and at the time of the interview all but one held senior positions.[25] Five were women. None were members of an ethnic minority, reflecting the fact that there were virtually no such persons occupying top positions in British criminal justice agencies. Nineteen of the Practitioners were graduates, with five holding higher degrees, including one of the nongraduates. The majority commenced their careers between 1965 and 1975, and one-third of them did so during the five years 1965-9. Their ages ranged from 40 to 65, but they were mostly in their late forties or early fifties. Most were, therefore, of the generation born during or shortly after the Second World War. Although their teenage years and early twenties occurred during the 1960s, most of them did not self-consciously emerge from the interviews as being of the 'sixties generation'.[26] However, it seems likely that something of the 'new consciousness' of that period played a part in their formative years.[27] Regardless of generational influences of this sort, the subjects' careers were touched by common events and professional fashions concerning crime and punishment.

Most interviews took place in the Practitioner's own office, and when this was not possible a mutually convenient setting was chosen. The interviews were tape-recorded and varied in duration from ninety minutes to five hours, with most lasting about two hours. The purpose was to generate a narrative that explored the origins, development, and effects of the subject's working credo. The issues explored included formative influences, turning-points, education and professional training, crises, and support structures. Questions were also asked about reform tactics, including problems of changing organizational structures and traditions. I adopted an appreciative style, appropriate to a conversation. The approach resembled the informal interviewing procedure, described by the American psychologist Jerome Bruner as being 'designed to encourage meaning-making by narrative recounting rather than the more categorical responses one obtains in standard interviews'.[28] Bruner has no illusions that the interviewer can be neutral during the interview, holding that the 'study of a life', as told to a particular person, is, in some sense, a joint product of the teller and the told. The overall interview process produced what might be called a professional autobiographical narrative. The extracts that I selected from the narrative were reviewed and approved for publication by the Practitioners. In most instances the text remains unaltered, apart from grammatical corrections and clarifications of meaning. A few deletions have been made so as to ensure anonymity, and, for publication purposes, a small number of passages were amended so as to soften remarks made in the conversational setting of the interview. For similar reasons, three Practitioners deleted fairly substantial passages. Taken as a whole, however, the extracts presented here closely follow the original narratives.

The selected items address themes that arise with reference to working ideologies, and in particular explore the implications of Credo Three for criminal justice. The selection of extracts from a narrative carries with it several potential pitfalls. While every effort has been made to retain the narrative's context, there is the danger, as David Downes has warned, that one may miss 'the nuances, qualifications, and genuine ambiguities' that a fuller narrative might reveal.[29] A further problem is failure to capture fully what Michael Billig refers to as 'the dilemmatic aspects of ideology'. He insists that ideology be seen not as a complete, unified system of beliefs which tells the individual how to react, feel, and think, but as a mix of contrary themes.[30] There is also the possibility that an unduly static impression of the individual's working credo may be given. As Donald Polkingthorne has observed, 'Self . . . is not a static thing or substance, but a configuration of personal events into an historical unity which includes not only what has been but also anticipates what one will be.'[31]

Finally, the context might have been enriched by identifying the Practitioners by name; however, for professional reasons some of them requested that they not be named, and as a practical device, therefore, a code precedes each narrative item.[32] Where anonymity was requested, certain references to persons or places have been deleted. In some other instances, perhaps, the task of decoding will not daunt the *cognoscenti*.

None of the Practitioners articulated a precisely formulated working ideology. Rather, the general pattern was that aspects of the working credo emerged at various stages during the course of the narrative. As Walter Miller has suggested, credos tend to be preconscious rather than explicitly held.[33] While the degree of adherence to Credo Three varied, no instances emerged where the material cast serious doubt on the appropriateness of an individual's inclusion in the study. As Packer suggested with respect to his competing models of Crime Control and Due Process, both contain components that are demonstrably present some of the time in a practitioner's preferences regarding the criminal justice process. Packer correctly observed: 'A person who subscribed to all the values underlying one model to the exclusion of all the values underlying the other would rightly be viewed as a fanatic.'[34] The materials presented in this study strongly suggest that the relationship between values and a person's work must be regarded as a dynamic and developmental process. Personal maturity, professional experience, and the broader social and political context within which the individual lives are interwoven in the shaping of the working credo.

The interactive dimension of this developmental process is astutely described by A1, a senior police officer. This opening item also highlights two other general problems. First, there is the difficulty of influencing other people in the agency, a central theme in the study of complex organizations. The second problem alluded to here is the gap that often exists between the formal purposes and principles of the agency and the reality of day-to-day practice:

It is very difficult to talk about the relationship between personal philosophy and opportunity. They could be poles apart, but you have got to have your philosophy before you start using the opportunities. The development of your philosophy is denied you in a dominating organization, unless you are given the opportunities. The simple choice I had was either to be alienated within the organization and become extremely cynical, or clear out of it. I chose to clear out, albeit to another police organization. But, equally, it could have been not another police organization. It was being given that opportunity, and then being able to rub shoulders with people outside the police service who had a similar outlook, that allowed me to develop. It was an opportunity to develop. The move was absolutely crucial. It was that move and the opportunities that followed which gave me the confidence and ultimately the organizational clout (by being promoted) to bring about the change. Life is full of contradictions. When I was constable I thought, once I become a sergeant I will sort this job out, because this sergeant does not really understand what we should be aiming for. When I made sergeant, I felt very frustrated because the inspector was in my way, and I also found that those constables working with me would not always do what I wanted them to do. I found, particularly in large organizations, that the more power you seem to have in status terms, the less powerful you can become. Chief constables in large organizations appear to be very powerful, and in many senses are, but, in terms of affecting transactions on the street between constables and members of the public, they can actually be quite powerless. This is one of the major frustrations I find, in terms of bringing about organizational change. Having ideas and a philosophy is only half the battle, and in fact it is probably less than half the battle. The biggest problem is actually influencing people, getting people to identify with the philosophy and actually to implement it on a day-to-day basis. Perhaps some of our most notorious cases are where people have espoused the philosophy but are operating something totally different.

## CREDO ONE

The cluster of ideas, values, and sentiments associated with Credo One include a powerfully held dislike and moral condemnation of offenders, and the beliefs that as few fetters as possible be placed upon the authorities in the pursuit of criminals who, when caught, should be dealt with in ways that are punitive and degrading. Unsurpassed as a motto for Credo One is the observation of the late nineteenth-century English judge, James Fitzjames Stephen: 'I think it highly desirable that criminals should be hated, that the punishment inflicted on them should be so contrived as to give expression to that hatred.'[35]

In the early 1960s, D1 was a young officer working at an open young offenders' institution, at that time known as a Borstal. He describes a visit he made to Reading Borstal, which held young men who were either recalled to the Borstal system after release or had been disciplinary problems at other Borstals:

I went with an escort to take an absconder or some malcontent to Reading, which was a punishment Borstal. We walked around, and they were showing how it should be done, probably because we had come from an open establishment. An officer taking us round stopped at a door and said, 'Just watch this'. He opened the door and a kid dived to the door in order to stamp his feet to attention. The officer met him as he was trying to get to the door and he was pretty badly roughed up, while we were watching. That was put on for our sakes, and the boy was given a bloody good hiding. As is the wont of closed communities, it was not for me to say anything about it. We just accepted that this was happening at Reading, and that is why they went there. If you do not want some of that, then you do not go to Reading. This is rather simplistic when you are at the bottom of the pile, but I did think, Christ Almighty, what is going on?

There were times when I was personally involved. As 'Choky Block' officer we used to do punishment physical education in those days. Perhaps it reflected my own interest, but I drove some kids to the point of dropping with exhaustion. I think about it now and wonder why the bloody hell they allowed us to do it, but we certainly used to. We used to take them in the gym, and they would run and run and run. We had a great mound of coke and part of the punishment—we used to call it 'work'—was they had to fork this coke and throw it up to the top. With coke, you throw it up to the top and it just slashes down again. I was totally convinced that this useless work was the way a punishment block should be run. There was that sort of macho thing, which maybe has one or two positives, but mainly negatives. When it came to things like that, there was an element in me which believed in hardness and control. It was the acceptance of the new officer as to what was, and following what your colleagues have always done.

Twenty years later, as the newly appointed governor of a women's prison, he still encountered highly punitive attitudes, but by then he was able to directly confront what he found:

Staff attitudes were unbelievable, and like nothing I had ever met. I asked, 'Why do you behave like this towards women, and pregnant women and babies?' They ran the prison the way they thought men did it. I told them, 'But there isn't a male prison, certainly not an adult prison, where people behave the way you do.' Many of them were very good, but some seemed to go out of their way to try and make it as unpleasant as possible, to strip people of their dignity. There seemed to be nothing better than to have a naked woman jump up and down on the spot. I was absolutely appalled at some of it, and asked the Chief, 'Where did you do your training?' and she replied, 'What do you mean?' I said, 'Well was it Auschwitz or Belsen?'

Another prison governor, D3, took over a prison with a deeply imbued punitive ethos, but there were some indications that staff attitudes were beginning to change:

The average age of the staff is steadily coming down, and that will create a more healthy ethos in the place. For some, their attitudes to the job were formed anything up to twenty years ago. In that period the prison had a bloody awful job to do. It was the dustbin for the entire prison service, and they had some extremely difficult and violent prisoners to manage. It is no blame on them that those attitudes were acquired, but they are very difficult to shift. A lot of what we are doing as management will improve things when people can see that, if you treat people as rubbish, they will behave like rubbish.

There was a plan by some of the staff working in the vulnerable prisoners' unit to play a football match against some of their prisoners. That is completely unremarkable in most prisons, but here it represented a significant departure. Over the period between the plan becoming public and the match being scheduled, there were anonymous notes left on officers' keys, people were berated for being crawlers to management and 'nonce lovers', and so on.[36] In the event, some members of the staff team felt that they should withdraw from the football match and it did not take place. But it will happen, or something will happen, because it is the first time that it has got that far. Next time round they will actually screw up their courage and do it.

## CREDO TWO

The prevailing concern of Credo Two is to dispose of the tasks at hand as smoothly and efficiently as possible. The tenor is one of smooth management rather than of a moral mission. The description by James Jacobs of the initiation of a more professionally managed regime at Stateville prison, Illinois, captures the essence of Credo Two. The appointment in the mid-1970s of a new warden (or governor, to use the British term), David Brierton,

brought to the prison a commitment to scientific management rather than to any correctional ideology . . . Brierton is neither in favour of nor opposed to rehabilitation programs. His primary commitment is to running a safe, clean, program-orientated institution which functions smoothly on a day-to-day basis and that is not in violation of code provisions, administrative regulations or court orders . . . Brierton has brought a new definition of administration to the prison. He stresses efficient and emotionally detached management. He has attempted to remove the affect attached to handling inmates.[37]

Three views of the court system point to sources of the pragmatic ethos that underlies Credo Two. As a civil servant in the Lord Chancellor's Department in the mid-1960s, E1 visited a large number of magistrates' courts. He notes how courts adjusted to the pressures for pragmatic expediency. B1, a chief crown prosecutor, outlines the process with respect to the prosecutor in the magistrates' court. Finally, E6, a barrister and recorder, describes the expression of Credo Two among the judiciary.

E1, a justices' clerk:

> Sometimes rituals were played out and devices used by key actors on the official side to deal with persons stepping out of line. Completing the task in hand has a high priority, with the risk that people might not have their 'day in court', in any true sense. There were tidy methods which those within the organization could use to outmanoeuvre anyone who might otherwise slow down the system unduly or obstruct it.
>
> If the standard is one of narrow attitudes, a procedure-based or mechanistic approach to what the job is about, of processing cases through the system, the organization can start to think in that way. The prevailing atmosphere can spill out and take a hold on every part of the organization. This is not to imply that anything underhand is going on, but people within the organization will find it difficult to resist the momentum. A heavy atmosphere can prevail, and this runs the risk of a predisposed attitude towards people coming before the court, who find themselves on the wrong side of a cultural divide.

B1, a chief crown prosecutor:

> The more you appear on a day-to-day basis in the local magistrates' court, the more cynical you become. It is inevitable that that should happen. All that you see in the court are people who have committed further offences while on bail, or any of the other grounds for which they can be remanded in custody. If you are not careful, you fall into the danger of becoming very prosecution-minded. It is very difficult at times for the prosecutor to be able to stand back. It is all very well saying to him,

'you are independent'; it is another thing to get him to stand back and look at the thing objectively, because the very nature of his job is that he sometimes inevitably adopts a cynical attitude to what he sees.

A leading barrister, in conversation, once referred to the judiciary as being 'drowned in the ethos'. E6, a recorder, had no difficulty recognizing and articulating this notion:

> The 'ethos' includes doing things the way they have always been done; a belief that our system of justice is the best. That may be true, but it does not mean it cannot be improved. A tendency to feel under unfair pressure if the judiciary is criticized for its attitude to sentencing; a tendency to believe that, because they have been sentencing people for years, they know exactly how to do it, and that they do not need to look beyond their own courtroom at the effect of their sentences, and what is the purpose of it? It is quite easy if you are a full-time judge, and quite difficult to avoid in fact, to become self-opinionated. If someone is bowing at you all the time you are going to start to feel rather important.

E6 further develops this point in relation to Lord Denning's remarks of 1980 with reference to the Birmingham Six, that to question what the police were saying might lead to the 'appalling vista' of undermining the public's confidence in them:[38]

> It was a disgraceful thing to say. It was in the context of the applications for legal aid to pursue the civil action when they alleged they had been beaten up. Those men were beaten up by somebody, and the argument was whether it was the police or prison warders. To say that was exactly typical of that generation, and behind it was, 'we think you are exaggerating; we don't think it can be as bad as this'. That is something which is engraved in a lot of the senior judiciary. It comes from having been brought up in an environment where you accept that what the police say is true, because the alternative is pretty horrific.

A more extreme picture of the criminal process is drawn by E4, who has represented many persons charged with terrorist cases, where the pressures to cut corners may be especially powerful:

> As long as you can caricature that person as harmless to yourself by making them appear thick and stupid, you do not have

to care about whether they have got legal rights or not, because they have ceased to be human beings. They have ceased to be anybody that you are afraid of and might accord respect to. Reduced to a caricature, they are worthless, and all you can do is just simply throw them away into a heap called a prison. It is very difficult to start standing up and saying to people, 'I am sorry but the system requires a fair trial and we have the obligation to give not the semblance of a fair trial, but the fact of a fair trial.' I still have not managed to convince some of Her Majesty's judges about that.

I had a man who was doing twenty years and had not been shown a single sheet of paper in the depositions of the case against him because his solicitor took the view that these were covered by 'security'. How on earth do you take instructions on a case if you do not allow the guy to read what forms the evidence against him? That had been done by a solicitor who would have got extremely upset if anybody had suggested that he was behaving improperly. Yet stepping back from it now, I doubt if anybody would be prepared to accept that his behaviour was proper. He had become intimidated by the crime of the man that he was representing and therefore had ceased to be a solicitor. He had become just another animal in the herd, and as the herd instinct took him, so he reacted to other people's fears.

E4 became embroiled in a long-running dispute with Home Office lawyers arising from efforts to establish that prisoners had basic rights. He describes the blanket denial that any such rights existed as 'an astonishing example of government pragmatism'; it was an easier course to argue that prisoners had no rights than to attempt to establish what these might be. He also remarks on the gap that may exist between a person's beliefs and actions:

It was a kind of dishonesty, which used to make me so angry because I did not believe that these people were green men on another planet. They were people I should be capable of respecting and working with; they lived in decent houses and had decent standards; they had children who were clean and well brought up and did not throw food around the table; they prided themselves on their intellectual capabilities; they came from the best schools and universities—and yet they could come

out with this arse-hole stuff. I have never had an explanation for that. I do not understand what makes people do that
because at the end of the day it is a negation of everything they
think they stand for. And yet they do it: why? That is the great
unanswered question of the day: why do they do it?

This theme is developed by D1, a prison governor, in the context
of a senior colleague's betrayal of values which he believed they
had both shared:

He and I finally parted company in terms of sharing a belief
system. After the riot was all over, we had a whole load of staff
come in from another prison. I went down to Board of Visitors'
adjudication and they carried this unfortunate prisoner in, and
he had really been battered. I said, 'What the bloody hell's
going on here?' They supported him in and he was charged with
assaulting staff, and when the staff let go of him he just slumped
to the floor. He had been tanked. The adjudication carried on,
but by then of course the Board of Visitors were all caught up in
it. I thought, somebody is going to notice that he is not very
well. Nobody noticed at all, and they took ninety days away. I
said, 'Could I ask the prisoner a question?' They thought I was
going to ask a technical one. 'Could I ask him how he came by
his obvious injuries?' He said, 'Oh I was beaten up by the staff.'
I said, 'Was it by our staff?' He said, 'No, it was the foreigners.'
I said, 'I'll see you in your cell.' So I went in and he said, 'We've
had our tear-up. You've had your bit of business. Let's all get
back to fucking normal.' He was the most sensible of all of us. I
told the staff that the next member of staff I found involved
would go down for five years, and from that minute it stopped.
I then rang the regional director, and was absolutely shocked.
(Maybe it happens when you find out that someone you had on
a pedestal is just human.) I rang him up and said that there had
been some naughtiness going on in the block, and he said 'I
know.' I asked, 'Well how do you know?' And he said, 'Both
the governor and the doctor have been on to me two or three
days ago and told me that it was going on.' I was the deputy
governor and I knew nothing at all about this. I had literally
walked in on it. I had to stop it in all conscience. This led to a
vote of no confidence from staff, but high credibility from
inmates, which stayed with me for years. I said, 'You're the

regional director. You know there's beatings going on, and neither you nor the governor's done anything about it.' I said perhaps more than I should have done, but I really thought that he was a man. He said, 'Well it is political, you have got to have your staff with you.' I said, 'No, no, no—you either accept that or you stop it and there's no half way.' This man had had this tremendous influence on me, and something had happened between us. It was a very great loss to me on a personal basis.

The issue of betrayal and the personal tensions that may arise are set forth in general terms by another prison governor, D4:

How often have we, in little things or in greater things, failed to take a stand? How often have we seen that our colleagues have failed to take a stand? How often have we gone along with it all and eventually been converted to the other way of thinking? By gesture or by silence, we have condoned or given the message that we will turn a blind eye when essentially we lack the guts to say no or to stop something from happening. We all know that it takes greater courage to say no than it does to say yes. We fail to realize how easy it is to become like those we oppose.

These extracts suggest that the threat to Credo Three is more likely to come from the pressures that favour the pragmatic and expedient stance to criminal justice than from the crude assertions of Credo One.

## CREDO THREE

The core principles of Credo Three are empathy with suspects, offenders, and the victims of crime, optimism that constructive work can be done with offenders, adherence to the rule of law so as to restrict state powers, and an insistence on open and accountable procedures.[39] The contemporary expressions of Credo Three include a minimalist view of criminal justice intervention, and an awareness of the connections across the criminal justice process and of the reform potential that derives from an appreciation of this interdependence. Credo Three also firmly locates criminal justice within the broad arena of social policy, and makes the linkages with such issues as housing and education.[40] Jerry Miller has captured the credo's essence:

The best model I could use is what I would insist upon doing for my own were they in trouble. If we use that as the measure we would be in quite good shape and one then moves into an entirely different realm. This idea that delinquents are somehow or other qualitatively different from the rest of us, or that our own have no propensity, possibility, or potential for involving themselves in similar or worse behaviour, is nonsense. The criminal justice system has had a long history and tradition of separating, of stereotyping, and of providing the definitions that allow us to exclude our fellow human beings from the human family.[41]

Jerry Miller's own working life provides a classic exemplification of the Credo Three practitioner. In the preface to his account of closing down the youth institutions in Massachusetts, he writes of the experience of sometimes, when listening to music, being overcome by a euphoria which makes it impossible to be still. Only by getting up and walking about could he ground himself:

Though I could be more moved by music than most anything, it was a rare accident to lose hold of time and place. I take it, it has to do with some unresolved adolescent regression—but, oh, what a one—and hardly to be relinquished. Why mention these things? Because they are the closest I come to describing the experiences which routinely overtook me as I tried to deal with the problems in the Department of Youth Services. They inevitably rose to my consciousness—tied to a face, a particular youngster, a haunting story which kept me going, occasionally transported me, and always gave matters meaning.[42]

There is another great practitioner, also an American, whose lifework epitomized the core values of Credo Three. Thurgood Marshall was chief counsel to NAACP, a judge on the US Court of Appeals, Solicitor General of the United States, and for twenty-four years, until his retirement at the age of 84 in 1991, a Supreme Court Justice. For Justice Marshall, a former law clerk recalled, 'it was worth fighting with his colleagues when they ignored the temptations government officials often have to deceive in order to achieve their goals. He fought to fulfil the Court's duty to guard against those failings. This mission meant telling and facing up to the often dark truth about government. It meant not forgetting that government employees are human, too; they can sometimes lie and use tricks contrary to the directives of the Constitution and the demands of fairness. Officials may shade their testimony to secure convictions. Given the chance, they may try to bypass constitutional restrictions and take advantage of the pyschological vulnerabilities of people they suspect of committing crimes ... He insisted that shameful episodes in judicial decision-making are not to be hidden from view, but remembered and recalled precisely when they are in danger of repetition.'[43] As Marshall's long-time ally on the Supreme Court, William Brennan, wrote: 'He has never stopped challenging us to make the Constitution fulfill its promises to all Americans; he has never stopped calling upon (in Lincoln's words) "the better angels of our nature".'[44]

For many of the Practitioners, the starting point is that of minimal intervention. The scope of the criminal law should be narrowly drawn, and when intervention is warranted it must be tightly circumscribed. This concern arises from recognizing the inherent capacity for damage that might be inflicted by the criminal justice apparatus.[45]

The minimalist proposition is boldly stated by A3, a chief constable:

> When the police are dealing with people in liberal democracy, the agenda should be attainment of objectives with minimum social casualties. The task is one of assisting society to move along with the least amount of damage to all concerned and the maximum amount of tranquillity commensurate with liberty.
>
> I want to know what are the principles of a tranquil social order.[46] If I can produce a tranquil social order I may not need policemen at all! The use of the police machine is the recognition of failure. The excessive use of the police machine is a failure of democracy itself. If you need a heavy police machine to keep the body politic in order, then democracy has failed. I would start my thinking policeman worrying about the social contract, because we used to argue about the social contract between the police and the public. I go back to John Locke rather than Thomas Hobbes. It is to reason why: that is what policemen should precisely do, to reason why.

C1, a chief probation officer, reflects upon the shift to a minimalist perspective among liberal practitioners, and describes how he was able to convince a probation department imbued with a strident interventionist tradition that it should change course. This item illustrates the extent to which the expression of a value position may alter over time. As discussed in the next chapter, the rehabilitation model adhered to by many liberal practitioners had by the early 1970s been largely abandoned:

For a long time, we felt we had a licenc
for the best of all possible reasons. Part
criminal justice itself has moved and pa
temper of the times, we have learnt that we
One of the first things I did in my present j
committee to redefine its objectives and prior.
off with a set of principles, the first of which
intervention. We demonstrated to the committee,
juvenile justice system, that non-intervention is a ᴖotent
force for good. That colours our attitude as to who we ought to
be involved with in a way that it never used to. It is clearly what
I have learnt most along the way.

D4, a prison governor:

My starting-point is not to be over-ambitious in terms of what
you can achieve. You can have tremendous ambitions in terms
of rehabilitation, almost suggesting that in some way you can
get rid of crime. But what is the point of trying to achieve the
impossible, because all you get is a lot of frustration and a
waste of energy and resources? It is better to look at prisons in
terms of where they are, and not where you think they should
be. This is highlighted, in particular, by the facts of three to a
cell, the primitive sanitation, and the 23 hours in cells. The
basic thing we have got to achieve, and this is my simple aim in
life, is to get good reasonable prison living conditions. If you
treat a man reasonably, you will get a reasonable response. My
arrogance is that I believe that I can make a difference. My aim
is that prisons should be places where there is no fear, and
where there is self-respect, dignity, and a sense of worth. That
would apply to staff as much as it applies to prisoners. It should
also be a well ordered community life. You have to accept
Michael Jenkins's comment that prisons cause more problems
than they receive.[47] Underlying the regime should be the simple
question: What are we doing? Why are we doing it? The typical
prison answer is that we have always done it that way. It is
amazing how many people work on that basis.

What we are trying to do is to make sure that there is dignity
and that we are presenting opportunities to people. Every
prisoner should have an opportunity to be involved in the
choices and decisions that are made about him. He should have
choices not only about where he works, but within the regime

nit that he is living in. What cannot be achieved, but
t we will pursue, is to try and replicate outside systems
inside. We still have walls and bars, but you can go down the
path of reducing the gap between the inside and outside world.

A related perspective is offered by another chief probation officer,
C2:

I saw the importance of connections in relation to community,
of the informal networks, and that has always stayed with me.
I realized that the criminal justice setting was an extremely
narrow, parochial, and dangerous setting for offenders. I was
imbued early on with a sense that the probation officers were
catalysts in terms of opening doors outside from the all-
enclosing criminal justice sector.

E1, a justices' clerk, emphasizes the exacting standards of natural
justice:

Most justices' clerks would recognize natural justice as the
guiding principle in relation to the legal and judicial side of their
work along with the main rules of procedure and evidence. But
differences of opinion exist concerning both the form and
the substance of natural justice. This is well demonstrated by
reports of appeal cases where courts stepped beyond the legal
line. Errors apart, there is a difference between being meticulous
and simply applying minimum standards, or applying the letter,
or interpreting legal principles to fit some desired result. It is
easy for issues of this kind to become dormant or to disappear
from sight altogether. If you keep repeating that natural justice
is important, people are reassured that it is alive and kicking.
Biased remarks do occur. The safeguard lies with those who are
prepared to challenge these. This will happen automatically if it
is embedded in the culture of the organization that high standards of natural justice really do matter. Repetition keeps the
principle alive and kicking. If it starts to disappear under the
mat, people wonder if it matters any more.

Another justices' clerk, E2, points to the need for empathy with
defendants in the administration of criminal law:

My advice to a new clerk or magistrate is be realistic about
what the system can achieve, because it is used by far too many,

particularly by government agencies, as a dustbin or as a means of resolving something that is incapable of being resolved. Accept that there is no real difference between the average defendant and the average professional, in as much as one in three people are going to be convicted of an indictable offence before they are thirty. As a clerk, remain true to your training as a lawyer. You are there to ensure that checks and balances are applied. There is a tremendous potential for the police to manipulate the evidence or for the courts to cut corners in order to get through the work.

I do not place myself as being any different from, by and large, anybody who comes before the courts as a defendant. I am perfectly capable of committing any of the crimes that they have committed. In terms of what the defendant is alleged to have done, unlike some of my colleagues, I do not feel on a different moral plane.

In sharp contrast to Credo Two, D3, a prison governor, insists that affect must play a part in the administration of prisons:

To do the job we do successfully, you have got to actually like prisoners. You have got to be prepared to accept that they will let you down. You then pick them up, dust them off and say, 'Have another go'. You have to be biased in their favour. That covers all sorts of things, but it can be exemplified with home leave. Every month I get many applications for home leave which have been sifted by a wing team and come to me with a recommendation. Nine times out of ten, the recommendation from the security department is that he should not get home leave, because, and I am exaggerating, many years ago he was involved in smoking cannabis. We should start from where prisoners are, which is that they are gravely disadvantaged and possibly difficult and disturbed, and try and move them forward. We should give prisoners better than a fair deal. It is no good imposing our own standards of what is acceptable and what is right, and somehow expecting them to conform and being punitive if they fail to conform.

Finally, A1, a senior police officer, spells out what openness should mean for the police:

One word sums it up—openness. All organizations are closed with quite clear boundaries. The police organization is not different from any other organization in that sense; but, because of the power it is able to exercise, so much greater is the need for openness. Openness in two ways: openness within the organization itself, that is, an honesty with genuine involvement and consultation, and passing of information at all levels within the organization; but more important, the openness between ourselves as an organization, as to what we are doing, trying to achieve, and why we are doing it. At the end of the day those people are our customers, some of whom may not necessarily recognize that they want to be our customers at the time that they are in contact with the police. It is an organization which has such power that it cannot afford to have secrets from the community. If there is a philosophy, and I am pushing hard with my colleagues and others who will shape the force of the future, it is to continue to move those boundaries in the organization. We have come some considerable way to keep those going further and further and further, so that secrecy within the organization is eradicated ultimately. That would ultimately be the objective, but is a long way off. By sharing, one will have that participation, because by opening up, other people are then able to influence us with alternative views. Even though one might never reach a consensus, the fact that one has entered into the process with some integrity has to be a good thing for the credibility of the organization. Regarding police misdeeds, this style and philosophy is absolutely crucial.

These extracts provide a way of looking at criminal justice that takes account of conflicting working ideologies. The narratives from which the extracts are drawn are those of practitioners who adhere to a set of liberal and humane values, and a common thread is the extent to which these values have to be continually reactivated and vigorously held on to in day-to-day practice. This means an insistence upon not taking matters for granted, of going beyond the formal statements of intent, and, repeating A1's words from the concluding extract, 'pushing hard with colleagues' in insisting that these values find and retain a place within the agency.

# 2 The Practice Context

THIS chapter seeks to place the Practitioners' working credo within the context of criminal justice in Britain during the period 1960–90. The development of professional ideologies can be fully understood only in the context of the broad criminal justice themes of the period. The principal themes that emerge from the Practitioners' narratives are the sea-change that occurred with respect to the notion of the rehabilitation of offenders; new approaches to policing, with particular reference to urban disorders; the mounting challenges confronting prisons; and, finally, the establishment in England and Wales of the Crown Prosecution Service.

It is not suggested that these four areas constitute a comprehensive overview of criminal justice developments in Britain during this period. They do, however, reflect the issues that emerge most fully in the interviews and provide a historical background for the material presented in subsequent chapters.[1]

## The Decline of the Rehabilitative Ideal

One of the most extraordinary and significant events pertaining to criminal justice this century was the rapid fall from favour of the rehabilitative ideal during the period 1960–75. This was the virtual collapse of the notion that the various sentencing options could be exploited for treatment purposes. As Tony Bottoms has written: 'By about 1960, there was a strong liberal consensus of informed penal thought in Britain, which believed that rapid strides would soon be made scientifically towards the identification of specific types of effective treatment for specific types of offender.'[2] Within fifteen years this consensus had all but vanished, leaving in its wake a clamour for more punishment, the ambiguous banner of 'just deserts', and a moral vacuum in terms of how offenders might best be dealt with. Nils Christie, although long wary of the excesses and abuses sometimes associated with the treatment ideology, has argued that ways must be found to preserve 'its own vital, but often hidden, message of compassion, relief, care and goodness . . .'[3] Rehabilitation was important as an ideal, in part because it provided a focal point for liberal and humanitarian values. The philosophical void that followed its decline allowed a free rein to proposals for deterrence, selective incapacitation, and a tougher 'no nonsense' approach to offenders.[4] As Christie observed, it became easier to 'inflict pain that is intended to be pain, and we do so with a clear conscience'.[5]

Three closely related circumstances largely account for the loss of faith in the rehabilitative ideal. First, a series of studies raised considerable doubts as to the efficacy of rehabilitative programmes to reduce levels of recidivism. The best known of these was the publication in 1974 of Robert Martinson's 'nothing works' article, and important questions arise as to why it made such a particular and lasting impact, and why his partial retraction five years later attracted virtually no attention.[6] The second circumstance was the mounting concern, mostly among liberals, about the lack of fairness afforded to sentencing and parole decisions that were shaped by rehabilitative considerations.[7] The third circumstance was the covert usurpation of humane values by some of the traditional opponents to rehabilitation. As David Garland has argued, while the notion of rehabilitation involves values such as care and compassion, the idiom preferred by the professionals was the 'bloodless language of social science'. When the collapse came, the caring values were a casualty because they had not become 'solidly entrenched in public attitudes or in penal policy'.[8]

The surest way forward is to pursue explicitly the expression of these values rather than to tie their fate to that of the rehabilitative ideal. Moreover, the particular mode of treatment may be much less significant than what people actually experience. R. D. Laing made this point in his inimitable style: 'The way that we treat each other is not reified as pills and things as treatment, but the treatment that we give someone is the way we treat that person; it should not be a noun, it should be an active verb. The way we treat one another is the therapy.'[9] The treatment idea, as David Garland has suggested, is 'only one—rather discredited—version of a much broader tradition of constructive penologies'.[10] But, while constructive activities are offered to offenders, eager intervention has given way to more modest expectations.

E3, a senior official at the Home Office, recalls the Home Office of the early 1960s:

> The 1959 White Paper, *Penal Practice in a Changing Society*, came at a time when the Office was at its most confident. It was the high point of the treatment model, the belief that crime could be dealt with by the way in which offenders were treated.[11] Ministers' speeches during that period showed a coherent and confident view of the solution to crime; police showed a coherent and confident view of the solution to crime: police technology to increase the detection rate, and careful assessment of defendants in remand and assessment centres like Risley. It was the period of the Streatfeild report, of belief in social inquiry reports, and of confidence in the effectiveness of carefully judged sentences which would send offenders either to probation programmes or to penal institutions.[12] Both would provide constructive regimes to deal with the diagnosis of the offender's problems. All that would lead to a law-abiding life afterwards and the gradual disappearance of crime.
>
> As we know, this did not happen. The seeds of disillusion were already there. Mannheim and Wilkins could be seen as the starting-point of the view that, whatever was done to an offender, it might not make much difference to his behaviour;[13] and, eventually, much later, to the view that measures to deal successfully with crime have to start much earlier than the point at which people are arrested and come to court.

D2, a governor, whose Prison Service career began in the early seventies:

> When I joined the Prison Service, rather naïvely, and like a lot of us at that time, I had a firm belief in the treatment model— that crime was somehow like an illness, and that if you found a cure then you could cure the patient. I began to see the injustice of the treatment model when treatment led to longer incarceration than would otherwise have been necessary. The Borstal sentence was indeterminate, between six months and two years, but such was the pressure of the Borstal system that a target date was set at thirty-two weeks. Nationally, Borstals were working to this target, except Lowdham Grange, where the governor believed that the optimum period of treatment was

seventeen months. Young offenders who ended up at Lowdham Grange with the treatment model got seventeen months, whereas elsewhere they got thirty-two weeks. It was from that I came more towards a justice model of containment, believing that people should be dealt with as adults. Rather than doing things for them, staff should provide facilities for them to help themselves.

D5, a prison governor, on his experience as a Borstal assistant governor:

The governor at the time was very much a 'Borstal governor'. He had been recruited as a housemaster at the end of the war and spent most of his time in the service working with the under-21-year-olds. There were two or three governors then of that calibre who were the remnants of the Paterson era.[14] That governor was very much in the rehabilitative mould, very committed to external activities, to camps, to encouraging the lads to become involved in positive experiences. That type of approach was a formative experience for me. Later I had to come to terms with how to transfer those principles out of the rehabilitative model and to express them in a more modern terminology. The traditional expression would probably have been that, if sufficient positive experiences are provided and a structure for these lads is set up, then they will not commit crimes in the future. In later years I had to work through the fact that that apparently did not work. But a justification still had to be found for dealing in a decent way and as humanely as possible with prisoners.

The prison is a very simple sort of institution. It consists of one group of people who are depriving another group of people of their liberty. To that extent it is a very stark environment. But, in order for both groups to survive, the task of containment has to be undertaken with a degree of humanity and decency. If I had thought at an early stage that it could not be done with a degree of humanity and decency, then I would not have lasted very long. I have always seen the essence of imprisonment in fairly simple terms, but the application of that simple principle is very complex, because it is about people and their relationships. What I saw with the governor was something that I could relate to. It was basically the notion that the individual, whether officer or prisoner, is a human being. Our situations are different, but our humanity is common. Frequently the application of that humanity was not very evident, but the principle that it was possible was evident. To encourage that and to foster it: that was the challenge of the job. The recognition that this principle was what we were about was always there.

At that stage I was becoming aware of the 'nothing works' debate. There was no encouragement from within the system to become involved in that debate. Nor was there discouragement. It simply was that it existed in another sphere. It was not seen as relevant to the practicalities of the system. I wanted to be stretched intellectually, and was beginning to explore and discover this literature. To that extent it was beginning to inform my own thinking, but it was not, even at that stage, informing the prison service. We were still living in our own fairly closed world of 'if only they do what we want them to do and become what we want them to become, they will be better people'.

C1, a chief probation officer, provides a frank personal reassessment:

It is quite difficult to look back, except to know that I have changed. It is partly my development, and it is partly the flavour of the times. By the time I left my first post, I was confident, probably over-confident, about what I could do professionally. I was an interventionist, to be honest, in a way that I am slightly appalled at now, because we did a lot of things then for extremely good motives that did not have a very good ending, in terms of taking people up the tariff quickly and getting them fairly swiftly into the criminal justice net. We did it for the best possible motives because we knew we had something fairly positive to offer. Individually a lot of those experiences were very good. With hindsight, I regret that we were so ready to intervene in people's lives. That phase was one of enormous confidence and a belief in a value of some of the positive things we were doing, but without the caution that comes with age and experience and about who you do it to.

We never bought the medical model as much as people thought we did. We did not have the post-sixties research which would have made us a good deal more cautious about global solutions than we were. I do not buy the 'nothing works' syndrome, but Stephen Brody's work,[15] in particular, made me think long and hard about who you would use it for.

C4, a chief probation officer, offers a post-modernistic view:

Some academics and researchers may have done themselves a disservice in the late sixties and through the seventies. The rather coarse criteria for effectiveness were flawed, and there was a more complicated explanation for all of this. A lot of practitioners were saying: 'This cannot have been all bad and all wrong. Let's look at it again.' After Martinson's 'nothing works', Gendreau and Ross is gradually filtering through.[16] I have always had a commitment that there was something positive in one's life work and my last twenty years could not have all been rubbish.

## Policing and Urban Disorders

Despite the Police Act 1964, which largely reflected the recommendations of the Royal Commission that reported in 1962, concerns about the accountability of the police have never been far from the surface.[17] The balance struck in 1964 between local and central control (from which the Metropolitan Police was excluded) has not been easily translated into practice, and there has been an uneasy trend towards more centralized control exerted by the Home Office, in part through the Association of Chief Officers of Police (ACPO).

A related concern arose from the increasingly pluralistic nature of the society to be policed. For this the police were ill-prepared, and the urban disorders of the early 1980s provided a dramatic reminder of the complexity of the challenge to orthodox police approaches. Lord Scarman's report on the Brixton disturbances brilliantly charted a policy framework that encompassed consultation with local communities, emphasizing consent rather than coercion.[18] These concerns were sharpened during the 1980s by the tumult associated with various industrial disputes, notably the events involving miners and print workers.[19]

Finally, there was the evaporation of public confidence in the police arising from a series of cases where huge doubts arose about the quality of evidence presented to the courts. In the early seventies, the Confait case led in turn to an inquiry, a royal commission, and eventually to legislation.[20] During the next decade, the cases of the Guildford Four, the Maguire Seven, and the Birmingham Six ensured that this issue remained in the public eye.[21] These anxieties reached well beyond the police, culminating with the appointment in 1991 of a new royal commission.[22]

A4, a senior police officer:

> In my early days in the police I had no organized philosophy, but an emotional feeling that underdogs needed all the sympathy and help they could get, a rebellion against the big battalions and those who had power and authority. In those days I did not particularly see the police service as having power and authority, because the police culture was very much more the working-class culture. In some ways there was an egalitarianism with the people we dealt with. They turned to us for their own problems or if they had disasters. We were on the side of the underdogs, and at least there was a feeling that we worked on equal terms with the people who were living in awful conditions. For example, police officers' conditions were often very awful, with young men living in sort of cattle stalls with three-quarter partitions between the beds.

A3, a chief constable on responses by the police to public disorders:

> In 1968 there were disturbances against the Vietnam War and a siege of the American Embassy. I had to train the police in new public order techniques. But we never used any hardware, and there was no question of issuing the police with bigger truncheons, crash helmets, and Perspex shields. We just put men in ordinary uniform up against crowds; and, although there was a lot of pushing and shoving and an occasional serious injury, in the main it was a triumph of minimum force. Those were the halcyon days which do not seem to exist so much these times. Society was changing then; the plural and the multiracial society was emerging which had a huge impact on the police in the sixties and drove the police much more into their own laager. The police began to feel that they did not understand the public and that society was changing for the worse. The Police Federation described them as the thin blue line. They saw

themselves as standing between total anarchy and a control of law and order which only the police could maintain. I always thought that was faulty thinking, and of course if the people had wanted to rebel they could have overwhelmed the police in no time. It is only because people want order that you have order. These were important thoughts, and still need to be considered and talked about.

A5, a senior police officer, on completing the inspectors' course at Bramshill:

I came back as an inspector and went to Brixton, which in the middle to late seventies was quite an experience. It became very obvious to me that things were going dreadfully wrong and wanted adjustment—in terms of the police, but also in the way in which the community was prepared to influence certain levels of behaviour. I was worried about it then. (It looks easy now.) Brixton was like a time bomb, and we had some responsibility for the way in which it was policed. It was also about people not feeling that they had a stake in the system.

C5, a chief probation officer, recalls working in Brixton during the early 1970s:

I wanted to work in a multicultural area, and chose the Brixton office. It had a big impact, raising issues about racial harmony and the discrimination that people face from different groups. It was a very pressurized time, and there were quite high case-loads and dealing with some very disadvantaged people. It was very good groundwork for me later. We certainly addressed racial issues as a local team. There were a lot of problems and tensions with the police at that time, and quite a lot of abuse by the police, which we were very conscious of, and tried to work within. I am very conscious of these now, in terms of anti-racist practice. Looking back at my social inquiry reports, we had really been quite racist, and had labelled people quite badly. We have since developed our understanding about what influences sentencing in terms of racism, in ways that we were not aware of then. We have tried to tackle it, and to build up very good links with the community, which I think we have done with some success.

It was pretty dreadful in Brixton at that time, as was recognized by Lord Scarman following the riots. There were terrific numbers of people going to court under the 'Sus Law'.[23] On some contacts I had with the police, they said, regarding a client of mine: 'Well if you tell him to plead guilty I'll make sure he gets bail', when he clearly was not guilty. We had quite a few clients (we used to be opposite the police station) come over covered in bruises, claiming the police had done it. All we could do was give them advice about how to make complaints through the proper procedures. We did not work well with the police, because we were miles apart from them. That is no longer true, but it certainly was then.

The criminal justice environment is one of flux and change. Although predicting the direction of change is fraught with difficulty, there does sometimes appear to be an oscillatory process at work. Progress is made on one particular front only for retrenchment to follow in due course.[24] This 'see-saw' pattern of criminal justice reflects, at least in part, the ebbs and flows in the strength of competing values.

A1, a senior police officer, describes one such oscillatory pattern with reference to the police:

In traditional terms, what matters to the organization is hard-line detecting criminals and chasing crime. Anything that waters that down is ancillary. There is no doubt within the police organizations around the country that there is a move towards a service orientation. It is a gradual process and one which has not been completed. However, it seems to be sustained only so long as certain individuals are prepared to support it. It tends to be responsive to its environment, in the sense that, when things become difficult—with public disorder, riots, and so forth—all the progress that has been made towards a service orientation tends to take a step back. We all tool up—and it is plastic bullets, CS gas, and all the rest of it. It gives those people who actually feel comfortable with that traditional force-type organization the opportunity to put the brakes on the soft approach. A major problem that any police organization faces is the ability to work up a philosophy and sustain it. It tries to do everything, but it does not really understand what it has tried to do. Until it gets to grips with that, any notion of police professionalism is pie in the sky.

## Prisons in Trouble

During this thirty-year period, the prison system was confronted with numerous related crises. During the mid-1960s, there were several much publicized and highly embarrassing escapes, culminating in 1966 with that of the spy George Blake, and the subsequent inquiry by Lord Mountbatten and a tightening of security across most prisons.[25] In the early 1970s attention shifted from security to control, in the wake of a series of peaceful demonstrations by prisoners as well as violent disturbances. These events, in turn, contributed to a very much more militant stance adopted by prison officers, and industrial relations remained volatile throughout much of the next two decades. The relentless increase in prison population size throughout most of this period had implications for every aspect of the prison system, particularly contributing to gross levels of crowding and a very poor range of activities for prisoners and appalling working conditions for staff.[26]

Finally, there was a lack of vision and purpose in the management of the prison service, which was especially evident in the wake of the disturbances at several prisons in April 1990, most notably Strangeways Prison in Manchester, which was virtually destroyed.[27]

There are significant differences in the recent prison history of Scotland and Northern Ireland. Serious prison disturbances occured in Scotland in the early 1980s. Although these incidents did not lead to judicial or other forms of independent inquiries, they prompted a far-reaching reassessment within the Scottish Office and prison system. In Northern Ireland, the prison system has been largely preoccupied with large numbers of people sentenced to very long terms of imprisonment as a result of Republican and 'loyalist' terrorist activity.

D6, a prison governor on his first posting to Wormwood Scrubs Prison:

> I went to Wormwood Scrubs in July 1966, and in October Blake escaped. I saw a quite dramatic transformation from what had been a very liberal prison over the years, with an open regime within the walls. Within a matter of weeks the whole thing was turned on its head, with dogs, cameras, lights, and

fences. Because of the preoccupation of senior management with the recovery of Blake, we were left to get on with keeping our bit of the prison afloat without any supervision and interest from senior management.

The impact of Blake's escape on the prison system dawned on me rather slowly, partly because I was very junior and inexperienced, and also because I was not right in the eye of the storm as the local prison was slightly to one side. Although I did not have the kind of perspective that I would have now, I did realize that something very dramatic was going on around me. At no notice, and through half the night, we had to go through every single prisoner's record. That became the identification of prisoners who were going to be Category A. It was right that certain things disappeared post-Blake, but we lost a lot of good things too. I became aware of the impact at a human level, and there was talk in the governor grades of course that the then governor had been shockingly treated and was retired about a fortnight before his retirement date. That threw everybody, both as a decision in its own right, and because there was an awareness that he was being made the scapegoat.

The implications for the management of prisons and the role of the prison officer are explored by D6 in the process of accounting for militancy of prison officers after the mid-seventies:

The understaffing of prisons in the sixties was quite incredible. But it had a positive side in that prison officers had to work with and through prisoners. There was a sense of camaraderie that was very attractive, [of] working together, and relationships with prisoners had that element, because both sides recognized that that was the only way to keep the thing going. There were inevitably some bad aspects to that, but there were also some strengths which were subsequently lost.

We had to re-establish managerial control of running the Service, and [of] the overtime issue in particular, which was bleeding the Prison Service and not creating the kind of prison officer that was my professional aspiration. In the 1970s the intrinsic job satisfaction of prison officers had been replaced by a preoccupation with earning money, by extrinsic rewards rather than intrinsic rewards. Things were going totally in the wrong direction. I was very clear that we had to turn this thing round somehow. While one might quarrel with some of the tactics, I was in tune with the strategy.

There were two broad reasons for prison officer militancy. One was internal and one external, and the two went together in the context of changes in society and the changing attitudes to authority. The internal reason was that the prison service began to reap the harvest that it had sown in the sixties, when we had exploited staff, perhaps without realizing it. It goes back to manning levels, and to assistant governors being trained primarily to deal with inmates, not with staff. There was no notion that you had to have staff management skills. Governors did not understand the staff attendance systems and left all of that to the chief officer. It was hardly surprising that in the end the prison officers blew the whistle and said they would no longer put up with it. That was our own fault, and, in the way that often happens with social change, the pendulum swung rapidly from one extreme to the other. This was coupled with the collapse of the medical model, resulting in a professional ideological vacuum. The Service was no longer clear what its purpose was or what it wanted staff and managers to do, which aggravated the position. Prison officers were left not knowing what their role was, and therefore began to turn to industrial weapons, with the preoccupation on mechanical issues such as how many staff there were, with numbers rather than quality of work, and with pay rather than job satisfaction.

The external reason was that the seventies saw a heavy recruitment of prison officers, with the main source of recruits moving away from the armed services, from people who had a background of discipline, to industry, particularly the mines. We found ourselves with people coming in with an industrial mentality, and with experience of trade unionism. These forces came together. The combination of the internal and the external factors produced an environment which management was not equipped or able to handle. The result was that management lost control, and the prison officers' trade union began to determine the way things should go.

D5, a prison governor, on prison disturbances in Scotland during the 1980s:

It was difficult to identify any one exclusive reason for the trouble. There were a variety of contributory factors. Overcrowding was significant. There was the Secretary of State's statement in 1983 about restrictions on parole.[28] There was clearly heavier sentencing by the courts, particularly of people who had committed drug-related offences. Very often, incidents in the prison system happen because everything conspires at one time and a spark sets off the incident. I was struck by the fact that the disturbances all occurred in large and fairly anonymous prisons, and that the small traditional town jails were not involved in these troubles. I had the fortunate experience of opening a new prison. It was a Victorian but refurbished prison and we managed to set a new tradition very successfully. We had long-term prisoners, some quite difficult former prisoners from Peterhead and other establishments like that. Despite that, there was no internal trouble or disruption within the prison. That confirmed all the premises under which I had been working about the size of establishments, the relationship between staff and prisoners, and humanity and decency.

## Establishment of the Crown Prosecution Service

The setting up in 1986 of the Crown Prosecution Service in England and Wales has had fundamental implications for the administration of criminal justice. The prosecutor is becoming a key player at virtually every stage of the process and may emerge as central as in many countries of continental Europe. In particular, the wide brief to interpret 'public interest' considerations enables crown prosecutors to exercise considerable influence on the shape and direction of criminal justice practice.

The Crown Prosecution Service had its origins in widespread concerns about miscarriages of justice and in campaigns by JUSTICE and other organizations for a new prosecution agency that was independent of the police. Paradoxically, this new-found independence coincided with an emerging appreciation of the potential benefits accruing from initiatives that took account of the interdependence of criminal justice agencies. This tension between independence and interdependence, and how best to strike the balance, extends across the criminal justice process.

B3, a chief crown prosecutor:

> The main drawback has been a period of uncertainty with a new department, uncertain of its new role and powers (and how it is going to exercise those powers), endeavouring to readjust its relationships with the other agencies in the criminal justice system in some areas of the country and to establish new relationships in other areas, particularly with the police. We needed to spend time looking at ourselves internally, but the whole thing moved so quickly that there was not time to do things a stage at a time. We were in being, and we had to be up and running and doing a job. The change from solicitor–client relationship to the present one, which I see as a partnership, has been delicate and sensitive, even in places where people are as sensible as they are in this county. The Crown Prosecution Service is still learning about the concept of 'the public interest'. We are trained as lawyers, and therefore the evidential criteria does not present many difficulties. But the public interest criteria are not straightforward, and it is sometimes difficult to ask young lawyers to suddenly start thinking in terms of philosophy, morality, public morale, and other nebulous concepts. It has to be done, and it is entirely right and proper that we are not obliged to enforce the letter of the law in every case but [can] look more widely as to whether cases need to be prosecuted in the public interest.

This chapter has sought to describe, mostly through the Practitioners' words, aspects of the criminal justice professional environment that existed between 1960 and 1990. The dynamic nature of the issues and discourses that constitute this environment means that transformations may occur over relatively short periods.[29] The decline of the rehabilitative ideal, for example, has had profound implications for the way many practitioners approach their work. More recently, the setting up of an independent prosecution agency in England and Wales has altered significantly the structural landscape of criminal justice. The development of working ideologies can be understood only in the context of these and other features of the professional environment. A recurring concern that emerges from the narratives is that of holding on to core values while adapting their practical expression to new understandings and changing circumstances.

# 3  Patterns of Early Development

THIS chapter addresses the formative influences on the Practitioners' working ideologies. As John Hogarth noted in his study of sentencers, the concept of self is formed by values, sentiments, and commitments. This professional self-concept is the key to how practitioners define situations within which they have to act. Hogarth concluded: 'The centre of the social space of a magistrate, therefore, is seen as his concept of self, expressed in his attitudes.'[1] The material presented through the following narratives does not do more than tentatively outline three developmental patterns of working ideologies among Credo Three practitioners. First, there is a pattern of consistent development where Credo Three was evident from the start of a career. Another pattern is of a distinct transformation of core values that might be likened to a road-to-Damascus conversion. Finally, there is a pattern of almost imperceptible growth and consolidation of Credo Three values without obvious turning-points or striking influences.

## Early Influences

Early family influences upon working ideologies emerged as being of particular significance in only a few instances. This finding probably reflects the interview approach, which did not deeply probe aspects of childhood development. The Practitioners were more inclined to identify influences they experienced as young adults and, in particular, to highlight the role played in this respect by particular persons.

From an early age, E1, a justices' clerk, was imbued with the values and expectations of his family:

There are two strands: one is to do with fairness, objectivity, and openness, and the other is about doing and achieving things. I was brought up in an atmosphere where both these were taken as read. They were important values through family, schooling, and the type of people that I associated with. You could do anything if you wanted to. It was just a question of setting about it. The other side of it was allowing for another person's point of view. It was absolutely critical that there would always be a favourable explanation to the most un-favourable situation rather than a worsening of a situation by people taking a critical stance. This is not a blindness to what is going on, but a willingness to allow that there may be two sides to the story. I grew up knowing I should not come to any conclusion until I heard both sides. The key to natural justice seems to lie in the existence and encouragement of these and comparable attitudes. No decision should be made until you have heard everything that everyone has got to say.

D1, a prison governor, grew up in a culture that condoned pilfer-ing and was deeply suspicious of authority figures:

My background was very low-level working class, born and brought up in an Irish working-class ghetto in Liverpool. My dad worked as a stoker in a factory, and my immediate family had no aspirations towards doing anything at all in this line. To this day, my mother is ashamed of the work I do. It is the only job she knows that is worse than being a policeman. That reflects our culture and our background. In my immediate family, there was a fair degree of criminality. There were people worse than us in as much as they were the ones who went to prison or to Borstal, but we were always at it. If anything fell off a lorry, we would be the first to be there, and when I was down on the docks, if there was anything that you could get out of the gate without being caught, you bloody well took it. But in connection with the work I have done, there is a certain morality which comes from Catholicism, which really came through the family. Getting a scholarship was an absolute disaster, because I went on to the then so-called grammar school and flunked out when I was 15 because with that sort of background it just was not possible to attain what I may have had. There were seven children, and the idea that one could do homework or study just was not there. I left school at 15 to work for the Blue Funnel Line. In our street, if people did not go to sea, they went to jail.

C3, a chief probation officer, was powerfully affected by a woman she met while still a student:

Nothing conscious had formed in my mind, but I went to university where history was my subject, and I was increasingly interested and drawn to the economic and social aspects of historical development through the eighteenth and nineteenth centuries, and obviously to the development of the philanthropic and humanitarian movements of the nineteenth century. I then had the opportunity for the first time in my life to go abroad, where I met someone who was a powerful influence on me—an administrator in a social welfare agency in Scandinavia. This was a woman with whom I felt quite a strong sense of rapport as a person, but it was also the way in which she talked about her job—she was nearing retirement by the time I met her—and some of what she had retained in terms of her early beliefs and motivations and passions, and what she wanted to do with her life to try and create the possibility of a little bit of change for the better. She was very influential, because when I went back to university for my final year I decided on a career in social work, and very likely in the field of criminal justice, or maladjustment of some kind. The contact was a purely social one, but because of the position that this woman held I had the opportunity, again quite by chance, to participate in a summer holiday camp for children with varying degrees of emotional disturbance and maladjustment who were staying on one of the beautiful islands in the Baltic, and I spent four or five days there alongside the staff, but not of the staff group, as a kind of participant outsider. I had literally no language in common with those children, some of whom were very difficult; they were early adolescents, the oldest was about 16, and there was a lot of aggressive and provocative behaviour. I had no language, and yet I found a rapport. With some of them I could communicate without words and have a calming influence, and I felt a positive response towards me. That was a very powerful experience of a non-verbal and non-intellectual kind.

Later, when I went for interviews, I talked about working with people, but inside myself it was about exploring those kinds of communication and rapport and influence that words help to channel and play an important part in. It seemed to me that I had discovered, in a way which felt very powerful to me, there was something beyond words as well.

For E6, a recorder, her interest in criminal justice issues coincided with a broader political consciousness:

I was quite taken with aspects of the criminal law, though I had no experience with it. In our upbringing there was an enormous respect for the police, and if a policeman walked up the street you felt that your respectability had been challenged. I had always taken an interest in court cases and was interested in criminal law from the point of view of the death sentence which still existed. I was teaching at a comprehensive school, and the head of the English Department was fanatically against the death penalty. There was a hanging, and he and I being terribly concerned about this had great rows in the teachers' common room. He got really uptight about it and put the humane view that there was no possible excuse. There was the horror of the papers that morning and the reports of scenes outside.

I was not particularly politically active. The first time I ever got concerned in anything was in 1956 when we were desperately demonstrating against Suez and pro-Hungary. That was a year when people started to think about things. I did not join the Labour Party until after university. That was not a family tradition, and my father voted for the Tories all his life.

In his late teens, C4, a chief probation officer, was attracted by socialist and humanist ideas:

When I was thinking of university, I had a voracious appetite for reading. Richard Hoggart's book, *The Uses of Literacy*, E. P. Thompson's historical work, and Raymond Williams's approach to culture and society fascinated me.[2] The movement at that time, if you were caught up in it, had a political, almost religious, commitment. Here was the truth of a new way of looking at things, which was very exciting to a late teenager wanting to do something 'meaningful'. Living in a mining community, I had seen deaths in families and people who got into trouble from the various working-class areas covered by the school that I went to; certainly not wanting to go down the mine, but equally wanting to engage in community matters, and why people got into difficulties and problems.

This thread goes back to my childhood, and all that reading and the interest in almost a socialist and humanist point of view is still with me. I had always had a rather working-class (if I dare put myself in that category now) disrespect for the Establishment. There is something wrong in society, and including the way criminal justice works; the Establishment are not quite as good as they make out.

I applied to join the probation service in Liverpool. They had about thirty vacancies, because nobody would go and work there, and I was the only applicant. There were about twenty-five magistrates on the committee, and on my application form I had put a line through where it said religion. During the interview every question, apart from one, was: Why had I put a line through that? Did I not know that the probation service started with the police court missionaries? Did I not know about the Protestant–Catholic divide in Liverpool? How was I going to cope with failure if I had no faith? The only other question that I was asked (and I was coming off an Applied Social Studies course) was, Had I ever heard of casework? They left me outside, and the only place I could sit down was a broom cupboard opposite the committee room. They came out at the end of the meeting and I had to catch the clerk and ask if they made a decision, and it became clear that they did not want me. But I was not going to put down that I was Church of England or whatever.

## Religious Influences

Religious beliefs were not a central issue for the group as a whole, although for a small minority these were especially important in their earlier years, and in some cases have remained so.

Christian convictions were, from the start, interwoven in E5's work as a leading magistrate:

My work as a magistrate was an outreach of Christian commitment. I had planned to go abroad as a medical missionary, but for various reasons this did not work out, and my concern as a doctor was essentially with people rather than disease processes. I was much more a social than a clinical doctor, and my work in the courts is a social involvement, but is quite distinct from being a social worker in court. It is interesting how religion affects one's views as a magistrate, because you are under judicial oath to do right to all manner of persons according to laws and the usages of the Realm, without fear or favour, affection or ill-will, and one's religious convictions should not enter into that. But they do enter it in that you believe in the possibility of redemption of any individual, however hopeless it may seem, and so as a Christian on the bench you will have a greater tendency to use alternatives that serve rehabilitation rather than punishment.

B2, a chief crown prosecutor, was offended by questionable commercial activities at the construction firm for which he worked as a lawyer:

The firm was aggressively competitive, and making money seemed to be their sole ambition. It was not that so much which offended me, but the dishonest way in which information was provided to auditors, and the desire to cut corners. As a lawyer, those things offended me gravely. It was largely a question of my being quite out of sympathy with their management style. Being a Christian, with a commitment to my faith, made it impossible for me to be party to anything which was unethical or dishonest. Being a lawyer also made it impossible for me to be party to anything that was unethical. I was there to service the group from the legal point of view, and not principally to make money.

As a practising Christian, I believe in forgiveness, and therefore I may take the view that a certain person need not be prosecuted because it was not in the public interest to prosecute someone who was entirely contrite, although that would depend obviously to some extent on the seriousness of the offence. Some areas of my work will be influenced by my beliefs, others are not. I wouldn't not prosecute someone just because they were contrite. Contrition might be the deciding factor in a small number of cases where the merits for and against a prosecution were evenly balanced. Other factors would obviously be relevant, such as the nature and seriousness of the offence, the attitude of the complainant and the police, whether compensation was a live issue, the age of the defendant, and so on. However, an admission of guilt, together with real sorrow, could be sufficient in some cases to justify a caution or other disposal. I draw a distinction between the influence of my Christian beliefs on decision-making of that nature, and my desire to give the prosecution service a bigger role in the criminal justice debate.

It is rather trite to say that's the way I am made, but I cannot think of a better way of expressing it. I have always taken an interest in matters of natural justice and in sociological issues. The wider criminal justice scene interests me just as much as the role of the prosecutor.

For B3, another chief crown prosecutor, Christianity became less crucial as a source of values:

Christian values were very much part of my family background and I had a Christian desire to see justice done in this world, at least in the part of the world that I could operate in. That has changed. I soon realized that justice has nothing really to do with Christianity. The ideals of youth have passed, and for me now it is an interest in people and how they come into conflict with the law, and how the law deals with very human situations, but not always in ways that lead to justice. People matter terrifically as individuals, and collectively as society. That influences the lawyer aspect of me, and the public interest criteria allows that to be. I could never be a person who could apply a set of rules indiscriminately without thought, and could not work in that sort of system. I have moved away from what I would see as the fairly restrictive Christian values to being a broader and slightly more political animal, but still holding that human life is important and that we have a duty to preserve the quality of life for the community as best we can. As I meet people and talk to them and see their perspective of life, I pick up little bits of them and I think, for example, that is a good thought: store it, develop it, and make it mean something for myself.

It is really maturity and involvement through life with human beings that has changed me. Being a parent changed me more than anything, and losing one's father or other people who matter makes you understand grief. It is important as a chief crown prosecutor to understand human feelings, frailties, and emotions, because you are not just administering a statute in the cold light of day: you are doing it in the melting-pot of society and human dimensions, and the public interest. The public interest must be people. How do people feel? Therefore it is important that you can understand people and what makes them tick. I hope I have developed as an understanding human being over the years, rather than being dogmatically akin to a set of established values or principles.

## Taking on the Challenge

Some Practitioners were drawn into criminal justice having been exposed to an aspect or incident which shocked them. While for others chance or drift played a role, it was often the influence of particular people that was crucial.

Accounting for his change of career, C1, a chief probation officer, observed:

> It was partly boredom, in that I was a heating engineer and you can solve everything on the back of a cigarette packet and I had been doing it for ten years; partly because of my first brush with the criminal justice system. I was a youth leader at a rural, but quite large, youth club, and three of the lads at the youth club were remanded to a prison. I was furious and appalled at the impact it had on both them and their families. That was the first time that I had come up against the whole system. But the connection was not very swift. The incident with the lads who were remanded to prison was followed by five years more in the electricity industry before I decided that I wanted to do something different with my life. It then became apparent that there was a range of occupations that I was interested in, of which probation might be one. That was when the connection came, of actually wanting to do something about a bit that I had seen first hand.

For E4, a solicitor, there was an unexpected and dramatic exposure to shortcomings of the criminal justice process in the wake of a terrorist outrage:

> After working in my office one Saturday in December 1974, I went home and turned on the television to hear that a bomb had gone off in Guildford. My instant reaction was to hope it

was not my office, and the thought of having to pick up all those papers. There was then a lot in the newspapers and the town was very jumpy and they arrested a number of people. I had a telephone call from the magistrates' court asking if I wanted a legal aid certificate for one of the 'Guildford bombers' and I said it should go to one of the bigger people because a case of that magnitude would monopolize me. They rang back a couple of hours later saying nobody else in Guildford would take the legal aid certificate. I said that was ridiculous and we have got to do it if the legal aid certificate is on offer. I rang the police station and said that I had been offered the legal aid certificate and wanted to see my client. There was a hostile atmosphere on the telephone and they said they would ring back. Eventually a detective sergeant rang and said I could come down the next morning.

D1, a prison governor, had been offered a job as guard in a Canadian jail by an unemployment office:

I resisted the idea, went home and talked it over with my wife, and then realized it was about twenty dollars a week more than I was getting on the unemployment, and that seemed a good enough reason to try it. I went back into uniform, and I suppose back into institutions, and started as a guard. When I first went into the jail, I was fairly happy with the rather distant custodial role that the guards had in a typical county jail in Canada, with no suggestion at all of any influencing of people. I was literally introduced to it on nights, and in a cupboard there were some Howard League books. I used to sit and read them, and maybe they fitted in with my own sort of concept of the way people ought to be treated, but they were the first seeds of a liberal idea towards prisons. It just seemed eminently sensible to treat people in a half reasonable way, and certainly the concept of treatment and relationships that came through then enabled me to start talking to lots of the inmates.

There was also the influence of the California system at the time. I read some of the literature about Scudder and the first open prison there.[3] Ever since then I have always leaned towards something of the liberal view which in many ways I see as treating people with a degree of humanity and dignity most of the time. It has paid off in relating to people and how one is seen as heading up an organization. The way to behave yourself and the values you show are often picked up quicker by inmates than they are by your staff, because inmates invariably see it as genuine concern, whereas staff tend to see it as being soft or over-zealous or even anti-staff.

D5, another prison governor, responded to a newspaper advertisement:

Initially, I had no burning desire to work in the prison service. I was looking for an opportunity to continue working with people and responded to an advertisement for assistant governors which I noticed in a newspaper. It was as simple as that. I really had no prior knowledge, feeling, or understanding for the prison service. The job of assistant governor seemed to me to be a people-focused job. People who knew me in my early life have difficulty envisaging me as a prison governor.

D6 was introduced to prison service as a student on a fieldwork placement:

The Prison Service had never crossed my mind as a possible career. It is not the kind of thing that you grow up wanting to be, like an engine driver or a doctor. The placement at Leyhill Prison took me very much by surprise, and there was no suggestion of a career aspect at that stage. It was very much part of learning about an aspect of the social and public service within criminal justice. As so often happens, the influences are the people rather than the organizations. David Hewlings, the governor, was trying to develop Leyhill into what would then have been called a 'therapeutic community'. After a three-week observation period I came away thinking, 'if these are the kind of people who are in the Prison Service, and these are the kind of things they are trying to do, this is a possibility'.

With David Hewlings there was the transparent care and compassionate concern for people, combined with a professional interest in managing and developing a community. The contribution David Hewlings made to the Prison Service was that he was one of the first, if not the first, to see that, if there was a professional skill about being a prison governor, it was to do with a particular form of management. He was the first person to use the term 'management' properly in the prison service. It was evident that what he was trying to do at Leyhill was to build this community, to shape it, to develop structures and systems and staff attitudes, and to involve prisoners in the running of the community.[4] That sparked for me, because it was consistent with the values I brought from an educational background. It seemed consistent with trying to run a school, or some other kind of institution. That began to make some intellectual sense to me at university when I was served up with Goffman's *Asylums*.[5]

There was a lot of enthusiasm and dynamism around, but from the outset there was a managerial context which was unusual. When I joined the Prison Service and found the values to be much more those of conventional social work, I was able to see fairly quickly that the Leyhill experience was an unusual one, for its time. For example, David had a lot of interest in what was going on in psychiatric hospitals and the development of their communities and used, in a sense as a consultant, the superintendent from a psychiatric hospital. There was that dynamic of total institutions—how should we manage them and what can we learn from other institutions?

## Professional Education and Training

There were mixed reactions to the influences of initial professional training, but once again it is the personal and direct impression made by particular people that is the common thread running through many of the narratives.

A4, a senior police officer, recalls her reaction to the police training school:

Coming from a school where there was an assumption that one was expected to ask questions and discuss things and find out for yourself, I found it extremely difficult to deal with thirteen weeks of rote learning and not being expected to question or argue. The training took place in a Victorian block in Regent Street which was miserable, dark, and gloomy. I very much hated that, but I have a very stubborn streak and it was an enormous relief to go out to the division in the East End and actually start doing police work. It was great fun, like being abroad—a totally new world to me, very rough working-class cum dockland culture, prostitutes, seamen's hostels, and remnants of the docks.

A2, a chief constable, recognized some conflicting aspects of his personality during his initial training:

I reported, I thought perfectly, a simulated accident in the street where a window cleaner had fallen off his ladder, but lost marks because I had not mentioned the fact that it was an eighteen-rung ladder. I could not see why eighteen rungs was more important than just a ladder. There were little aspects like that which I found a minor irritant, but I was quite a conventional soul—if you scratch the surface of rebellion, you will often find underneath somebody who is very conventional. I have always been rebellious and swung against the system, but I recognize in myself quite a strong, contradictory conventionality.

D3, a prison governor, on his assistant governors' course:

In great measure, the influence of the [assistant governors'] course had to do with the other people I encountered. Until your own attitudes have been brought out and tested, you are not aware whether they are the same as or different from those of other people. I was for the first time with a group of people, many of whose attitudes were rather different from mine, and there were some passionate discussions about aspects of the job. My tutor, Michael Jenkins, is someone that I admire considerably for his personal qualities and his integrity. I have seen him once or twice in public situations, prepared to stand up and be counted with views that run in opposition to the currents of the day.[6]

D1, another prison governor, looks back to the staff course with positive feelings:

It was a bit like the Navy, and if you carve your life up into sections it was one that was important to me. It gave me the opportunity to look in depth at work that I was doing; it introduced me to concepts of justice, and to criminology and penology, which I had only had a smattering of through the odd book. For me it was full-time study where I was able to share experiences with people who were very different. I had the opportunity to study, and particularly to take part in discussions, and was able for the first time to be influenced by reading and by the tutors.

The man who had the greatest influence on me there was Norman Jepson, one of the academic staff from Leeds University. His lectures touched everything which, both intuitively and from my life's experience, I knew to be right. He stood out among all the tutors, infecting me with his enthusiasm for a belief system of justice, decency, and treating people the way one should. He brought it all together, more so than anyone else.[7]

C4, a chief probation officer, recalls his social work course at Sheffield University:

I set up a small group of about eight of us on the course which we called the Humanities Society, and got people along to talk about philosophy, post-war opportunities, and what was developing at that time.

For one of my placements I worked with probation officers in Sheffield who were incredibly committed and were working with some extraordinarily damaged families and individuals. The tradition of pacifists working with the Family Service Unit had got into the probation service in Sheffield. They were just inspirational. Sometimes they gave me a lift home at nine or ten at night, and we would still be talking about the problems of the day.

C3, another chief probation officer, did her social work training at the London School of Economics:

I felt a great warmth towards people like Richard Titmuss and David Donnison.[8] Some of that warmth emerged even more over the years. It felt quite rich at the time, but looking back I realized that they prepared me for that intellectual rigour. Holding on to the warmth is important, but continuing along the direction of discovering and keeping your antennae open to what is going on around you is the equipment with which you are sent out into the world. It has been very important for me to recognize that you do not complete a training course and get a piece of paper and then practice it all in a way that says: I was not trained to do this, so I am going to resist having to do it. You have got to keep open and in touch with contemporary change. In the last few years, with the speed of change that has been happening, and also that has been required of us, no wonder I can feel so strongly about those early influences upon my training and development.

C1, a chief probation officer, highlights the limitations of his initial training in terms of placing the working situation within an intellectual and philosophical context:

> I was in a group of entrants to the probation service who were fairly conscious of a bright stream of graduates coming in at the same time. Most of us had made a conscious second-career choice and assumed that if we became senior probation officers we would be well content. I did the Rainer House course but realized how limited it was. It was only a one-year training course, and, while it fitted you to survive day to day as a probation officer, it did not do a lot in terms of some of the large questions that were around. Thanks to my first probation area, I did an external diploma at London University, which gave me the theory that I needed.
>
> I was still doing the job that I had come to do, but I was much better informed about why and how I needed to do it in terms of the range of theories and the range of writing and experiences I had actually been able to notch up. In my first post, I did things by instinct in a way that I did not do in my second, because I had something more solid, in terms of obtaining an intellectual framework, against which to test what I was actually doing.

On the other hand, D5, a prison governor, found the staff course of only limited value:

> The prison service staff course taught me the nuts and bolts of the prison service and that really was what I needed to learn. But intellectually it was pretty superficial and undemanding, and probably that was what eventually encouraged me to go down the road of further study. I had very little understanding of what the prison service was about, where it had come from, and the environment within which it was meant to be working. I needed to understand that structure, and that was why I set out to increase my knowledge and made approaches to my local

university which subsequently accepted me as a postgraduate student. My research was on the organizational development of the prison service. That was a good move for me, because the fact that I have a foot in both the academic and the practitioner camps has made it easier for me to survive in the system. I doubt if I could have done so if I had kept both feet in the field of the practitioner. I have enjoyed the intellectual challenge of having a foot in both camps, having to justify my practice in an academic context, and having to justify academic thought in a practical context.

A1, a senior police officer, raises rather similar doubts with respect to the Police Staff College at Bramshill:

> Each time I went to Bramshill, I was a little bit disappointed. It reflected the traditional organization and was not dynamic enough. I keep my contacts in the academic world, where I enter into debates, and find these keep me sharpened and challenged by alternative thoughts and ideas. *Outsiders* by Becker was the kind of study which I found particularly useful, as was Howard Parker's *View from the Boys*.[9] I did not agree with the way Parker approached that particular project, but what he had to say was not necessarily a view that a policeman would ever normally get. Goffman has been especially influential as I am very interested in interpersonal relationships and the relationship of individuals with the organization, and I still take *Asylums* and *Presentation of Self* off the shelf now and again.[10] There is a great danger of being a little bit insular within the organization and not being tested from time to time. My outlook tends to be oriented outside-in rather than inside-out.
>
> A major problem of the police organization is that it tends to be very insular. It goes through a consultation process but may not necessarily listen. This occurs at times with its own staff and with the public. There is an element of 'I listen, but do I hear?' As a chief officer, if one is hoping to keep in touch with public opinion, what is required is to orientate oneself to the outside of the organization and be prepared to listen, and more importantly to act. Being out there and looking in has always been my stance, but it is not necessarily comfortable for the organization which prefers to stay within the comfort of its boundaries and its traditional 'we know best' thinking.

## Early Professional Development

That early impressions and experiences are very important in shaping the course of a professional career finds support in these narratives. A common thread is the distinct effort made by the Practitioners to review the nature of the relationship between their concept of self and their new professional identity. As a general rule, the Practitioners sought to resist pressures to become 'organizational people' and instead sought out opportunities to influence the definition of their role.

C2, a chief probation officer:

> In my first job I had a case load of never less than fifty and often considerably more, and there was one particular month in the late sixties when I did twenty-two social enquiry reports. Nowadays, colleagues moan if they do more than six or seven. I thought nothing of working three evenings a week until nine-thirty and part of a weekend. It was a very total commitment, with no distinction between the private and the public life in a number of ways. It was a time to fulfil young men's dreams, of trying to bring professional and personal vision to that particular task.
>
> We set up a centre for the socially isolated called 'Friendship Unlimited'. We advertised in surgeries, social security offices, and with probation and social services. It was about informal networks, and it was one of the most important things I ever became involved in. We thought that care-givers should be the people who come to the centre themselves. Relationships were reciprocal and there is, if you find it, a need to help in all kinds of people, even those who have had experiences of mental hospitals and prisons or child care establishments. We found a disused brothel which was given to us by the City Council and opened up six nights a week. There were no rules or structure. We had somebody who acted as the host and a long-stop who floated round from room to room and made sure that people who wanted to talk could talk. There were games and facilities of a very basic kind, but it was a cup of tea, a chance to meet, and for some marginalized and alienated people to actually feel that they belonged.

A4, a senior police officer:

What I liked about my first post in Stepney, for example, as opposed to being a probation officer, was that I was not responsible for someone else's life. I did deal with crises, but many of these were not of arresting people but had to do with neglect of children and people whose houses were flooded out. It was helping people in crisis, but with the advantage that one dealt with it in an eight-hour tour of duty and resolved the problem in some way, and could then forget about it. The thing that I did not find attractive about the idea of social work was being responsible for somebody's life for two or three years and having to build up a relationship with somebody with whom one might be pretty antipathetic. Later I became better at shouldering long-term responsibility for the lives of some of the people who worked for me, and who had all sorts of domestic crises or difficulties at work. In this respect, perhaps I have become more mature.

I became interested in the death penalty issue in my early years in the police service. I always used to be in the minority, and it was one of the areas in which I always felt obliged to nail my colours to the mast and say what I thought. It was still a live issue when I was on the inspectors' course, and I found myself opposing the death penalty. I would occasionally encounter police officers who were opposed to it, but it was clearly an area in which I held the minority viewpoint, as with other things like expecting people to have a greater awareness of what was happening in the world outside. Coming from a background where one read books all the time, I have always been an avid reader, so I have continued to educate myself. The *Guardian* helped to educate me, and being a *Guardian* reader is one of the things which has marked me out over the last thirty years.

A2, a chief constable, on his first posting as constable:

I was posted as a constable to Bow Street and was pitched into the heart of the West End. I had been a rural child, with no concept of large towns, let alone the metropolis, and there was a lot of adapting to do. We lived in grim circumstances and were housed in hostels called section houses. Mine was over a police station, on its top floor, where we were required to live in cubicles, rather than rooms, where every breath drawn or nightmare could be heard from the neighbouring cubicle. It was a miserable winter, and we had to wash our blue shirts, and put them up to dry in a place where they could not conceivably dry; they would drip to a state of lesser moistness and we would collect them later, still damp and with a light layer of dust on them. It was very disagreeable, and we were there for a number of months before we graduated to an approved hostel where we each had our own room and felt as if we had moved into the Ritz.

I fitted into the very strong existing culture: predominantly male, white, and authoritarian. If anybody had heard me speaking in the early sixties they would have thought, 'Here is a racist'. On one particular occasion I had been in the West End and had seen two black men chauffeured by a white man, and I came back to the canteen where, presumably gauging a predominant value, I was speaking quite loudly to anybody that wanted to hear about this strange phenomenon that I had seen. I was taken aside by a senior constable who expressed some surprise at this view coming from an educated man and apparently not terribly well linked to the brain that was driving it. I was made to feel quite uncomfortable. That was, for me, an important formative point. So the culture was driven at a number of different paces by opinion-formers within it. There was the other culture, exemplified one November 5th. We were told that we were to be 'briefed' by the chief inspector. I had some sense of what a briefing was, and it sounded good and positive. I waited with anticipation on parade for the chief inspector to arrive for the briefing—how we were to manage the central London Guy Fawkes celebrations. The chief inspector came in and he said, 'Gentlemen, I have come here to brief you. The object of the exercise is to get one more arrest than last year.' He then walked out. I laughed along with the peer group, but was terribly disappointed at having had an expectation built up which was then shattered.

E3, a senior Home Office official, on his initial impressions as a civil servant:

There seemed to be a turning-point in the sixties in the way the Office thought of itself and was thought of outside. It had been a fairly confident and perhaps complacent department during the fifties and up to that time. The cases of Carmen Bryan,

Chief Enaharo, and Dr Soblen attracted a great deal of public concern about the way in which the Home Office operated and dealt with cases, and the way in which the tradition of care and concern, and the sense of balance, which was part of what the Home Office tried to communicate to young men and women joining the Office, could go wrong under certain pressures and in certain circumstances. The Office was not quite the same again after that. Of course, it was not just the effect of those cases: it was also the changing attitude of the media and the changing attitudes to politics and public administration more generally. It was a period when nothing could be taken for granted, presentation became ever more important, and issues would sometimes need to be argued with ministers in a context where the advice might be unwelcome to them. It was always the job of the Department and its officials to present the case and make sure that all the issues were properly recognized. But we still thought of ourselves as performing functions and carrying out procedures—we did not, as we do now, see ourselves as personally responsible for achieving results. But I remember some wise advice from my earliest days in the Office which is still as relevant now as it was then. My principal in my first job said to me that people should be allowed to do what they want to do: if the Office wanted to stop them, the onus was on the Office to justify itself, not on the person concerned.

B1, a chief crown prosecutor, on his initial awareness of prosecutorial independence:

I was fortunate to go to an authority where there was a very experienced chief prosecuting solicitor who had worked in many police authorities. Prior to my starting there, he had established an excellent relationship with the chief constable. However, he was not the sort of man who would do as he was told if the chief constable disagreed with him. He was very independent, and that attitude was inculcated in all people who worked for him. From the very first day that I started prosecuting, I was brought up on the idea that you are an independent prosecutor, and you do not do something just because the chief constable says you do it. Prior to the Crown Prosecution Service, I had worked in three police authorities and had never reached a situation where I was being told 'you will do that'.

There were disagreements—there had to be, as part of an inevitable tension—but I never got myself in a situation where I was being told I had to do something.

At his first posting to an open Borstal, D2 was aware that there were useful lessons about authority arising from the institutional situation:

From an early point, and this has been important in my work right through the time in the prison service, I was aware that most of the young men that I dealt with in the Borstal had had disastrous relationships with authority in one way or another. If one could encourage or make it possible for them to relate to authority in a less confrontational way, then at least that would be achieving something. I have myself on occasion had quite prickly interactions with the police, albeit over petty things, but where I have found authority figures, in my view, needlessly sarcastic or provocative. Because, I suppose, I am middle-class, articulate, and polite, I have usually managed a calm response; a lot of my Borstal lads would not have been able to do that at all and would have responded, perhaps, with a punch. The other side of facilitating a reasonable relationship with authority figures is that it is incumbent upon the authority figure to be reasonable at all times. Something for which I always strive is that one should not be simply autocratic. One is an authority figure as a prison governor. If governors act unreasonably, why should prisoners follow our instructions or take our advice?

D1, a prison governor, recalls his initial experience as a prison officer:

I had a tremendous interest in trying to influence them [the young men in the Borstal] because, in many ways, I identified so many of them as being exactly the way I had been and indeed still was. I happened to believe that, for kids like that, all they needed was a break, like I had had. The idea that there was some innate criminality or delinquency in 99 per cent of them, certainly of those in Borstals, was nonsense. All that the kids needed was half a chance and most of them would grab it.

I easily fell into being able to relate to inmates both in terms of empathy and being streetwise as well. That probably goes back both to my own childhood and also to the Navy, because if you have to live pretty close to people then you develop certain survival skills including being able to read people.

D6, a prison governor, quickly appreciated that he would be most effective if he managed to work through prison officers:

A very strong influence was the rewarding elements of working closely with prison officers. I was able to form good and effective working relationships with the prison officer staff in general, and most crucially with the principal officers. That was, for me, what the job was about, because it was they who held the power. If you could not work through them and influence their thinking and behaviour, then really you were not going to be able to effectively do anything at all. This confirmed the impressions I had formed at Wakefield that, if one was going to do anything at all in the prison service, it was by working through the staff. My thoughts were moving steadily in the direction that this was a managerial job, not a social work job. I began more and more to question the value of the conventional social work training, to see that that was not the prime issue. It was important to have a grounding in some of the values about caring, but to train assistant governors as social case workers was not what was relevant.

D3, a prison governor on his posting to the special unit at C Wing at Parkhurst prison:

C Wing was a fascinating place, combining discipline staff and hospital staff. It was very much about dealing with individual prisoners and trying to produce a way of living and working with each individual which was going to keep him on a level keel. While I was at Parkhurst the impact of the cliché about whatever happens to prisoners happens through the prison officers was forcibly printed on my brain. One of the misperceptions that had been abroad, in the context of the staff course, and which my time in Borstal did not do much to change, was that assistant governors could be the deliverers of care, support, and therapy, almost in spite of prison officers. The big lesson from Parkhurst was that that was not the case: you have got to get the prison officers doing the right things and pointed in the right direction, otherwise anything you do will have no impact at all. We were able to expand some of the skills and motivation of staff there. The orientation of the whole unit was about caring for and managing difficult prisoners. You did not shove the bloke behind his door and tell him to fuck off: you actually had to deal with the situation. Its particular *frisson* was that if you got it wrong you were quickly in quite deep water. Staff meetings, which were often case conferences, were some of the most stimulating experiences I have ever had, where prison officers were talking in very sophisticated ways about prisoners. Part of what I value most about it was the experience of seeing what it is possible to achieve through staff.

I learnt a hell of a lot about prisoners, because we all spent a great deal of time listening to and talking to them. One of the bits of luggage I carried away from C Wing at Parkhurst was the ability to manage very aggressive people, which has been of immense value ever since. That may say something about me, but it also speaks about the system and the organization. At C Wing I was on my own, and, provided there were no major disasters, we could actually take the unit the way we wanted to. That is very liberating. It can also be disconcerting when nobody is talking to you about what you are doing. The most powerful influences were my own staff in the unit and the prisoners.

D1, a prison governor, on his first posting as an assistant governor:

I was proud that I had been sent there, seeing as it appeared to be so far ahead. There had been one or two articles by the governor about what it was trying to do, so I saw myself as being selected, rather than just being posted. I said to my wife, 'Aren't I lucky? They must think I'm good.' With these much older prisoners, I was very unsure of my ground. One of them, who was a Liverpool deadbeat, third PD,[11] alcoholic, and had spent most of his life in prison, called me to one side when I had been there about nine months and said, 'Look son, I like you, but you're making a right fuck-up of it.' I have been grateful to him ever since. He was found dead from alcoholism later.

The governor had been a bloody monkey on my back ever since I joined the service. He was quite an extraordinary person, and he allowed me so much rope to hang myself. The whole place with the old PDs was geared to treatment and involvement. He was trying to develop within a prison a thera-peutic community, and probably nobody has got closer to it than him. That suited my approach and belief system. We had this team of people who in many ways were a crowd of bloody cranks, but we all believed in the same crankiness, so in many ways it worked. We used to look at him with a degree of awe, because he could smile benignly, and we believed that he knew what he was trying to do. I accept now that he probably did not: he just had that bland exterior, which he could pull on when he was probably thinking, 'Where do I go from here?' He was breaking down the culture of the PDs, and some of them used to say, 'I just don't bloody believe this.' We had a council of prisoners that met with the governor every fortnight to talk about how to do it. We were doing things there twenty-five years ago which, if I could get them here now, I would be leading the whole of the prison service. We had groups on the wings and a personal officer scheme which I have never seen anything like in any other prison I have been in.

The governor continually exposed me to training courses and to experiences, like the Grubb Institute. He gave me very solid advice, sometimes advice I did not want, saying to me once: 'There is no place for you in the prison service as a governor grade because you are lower-working-class and it shows in everything you do and say.' At which stage I nearly went across the desk at him in a typical lower-working-class way. After about a month he called me in and said, 'Well you know what we've been talking about.' To enable me to grow, he sent me off on training courses, and if I am anything today, then it has got to be down to people like him in terms of maintaining any belief system that has come through the service, and I have managed to maintain.

For some Practitioners, defining their professional role was hastened by having to make an explicit choice between competing sets of values within the agency.

E1, a justices' clerk, on early conflicts he encountered within the magistrates' courts:

Important ideas came from reading books that were around in the magistrates' courts. These invariably emphasized natural justice, proper procedures, the need to observe statutory rules and regulations. Around the time that I joined the service there had been several appeals in these areas where courts had been criticized for acting wrongly, so that the topic was very much to the fore. They concerned such things as clerks retiring with the bench when they should not have done so, magistrates sitting in cases where they had some interest in the outcome, or bias being signalled by words or actions. A whole area of case law was developing. But on the courses I attended there was sometimes a double edge to what was being said. There was the rule as laid down by the High Court, then there were ways of getting round the rule or limiting its effect. For a long time, there seemed to be a tendency in magistrates' courts to look for ways of getting round what the High Court said, the justification being that the judges did not understand the practicalities, e.g. of dealing with truculent, often regular, customers, or a busy court schedule. This attitude was not untypical at that time. It struck me that the point of the exercise was not to get round the rules but to apply them in a decent spirit. If a natural justice rule said that the clerk should not retire with the magistrates, or dominate the proceedings, the purpose was to make sure that the magistrates were making the decision, openly, and being seen to do so. This represents a positive value, not something inconvenient that gets in the way.

E2, on his selection of one path rather than the one followed by the deputy justices' clerk, at the first court where he worked:

I blow no trumpets, but at the end of the day I think that is just me. I had a sense of unease in the way that he operated, dealt with people, and perceived his role. I was not comfortable in that role. I do not hold myself up to have any higher moral standing or legal abilities, but I am very uncomfortable dealing with people in that sort of way.

E5, on an early choice he faced as a magistrate:

For the first year I was a little bit ambivalent about all this. One magistrate who I thought was the exemplar of proper magisterial procedure, I later realized was a very biased sort of individual. He was a very upright man, but rigid and inflexible in his attitudes. A big influence was comparing the advice and attitudes of magistrates who had a more traditional approach with those of a more liberal approach.

I went in as rather a hard-liner. I felt that if you dealt with people fairly strenuously on the first occasion they might not come back. I was a bit for the idea that a fairly brisk reaction initially in court was likely to reduce the likelihood of re-offending. I had an exaggerated view of the influence of a few well chosen words from me in the juvenile court. Of the first twenty that we put on probation after I had become chairman of the juvenile panel, I discovered later that only one remembered anything at all that I had said. I changed my tactics fairly quickly. Over the years I lost much faith in deterrence in the magistrates' court. Deterrence may work for very sophisticated criminals such as drug importers and bank robbers who plan their escapades pretty carefully; possibly they work out the chance of getting caught and what happens to them if they are caught. But the vast majority of people in magistrates' courts are opportunistic in their offending, and deterrence has no real role.

The values that conflicted with those of the Practitioners, for the most part, fell within the Credo Two cluster. In the following extracts, A5, a senior police officer, recalls being confronted at the very start of his career with a set of values he was not prepared to accept. He makes the surprising observation that transfering to the CID enabled him to survive:[12]

My first impressions of other police officers, quite frankly, were quite discouraging. Some of the attitudes were such that one almost despaired about what one was doing. The police culture was something like that of the armed services. There was a sense of isolation, a degree of insularity, but there was a very close supportive spirit within it. Not always, but that spirit was as much to do with protecting the group as with providing a common bond of service. This had to do with the forties and fifties, and with a belief that if you had served your country then your country owed you something. There was also, what I have always seen as one of the bad qualities of professionalism, the notion that the professional is always right and the population as a whole—the potential client—should have very little influence over what professionals decide. There was the attitude of 'Who are they to tell us what to do?' It was a very poorly paid job at that time, and as a consequence large numbers of the people that joined were not of a particularly high quality in terms of their ability to actually analyse the objectives of what they were doing. These sorts of attitudes were slightly disconcerting to youngsters of fairly average intelligence, who increasingly were thinking about what they could look back to afterwards with satisfaction, knowing that they had achieved something.

Sometimes I got very close to leaving the police service. What kept me in was that, without even thinking about the objectives, the means were very interesting. I joined the CID, and if I had not become a detective I probably would not have stayed. Although there is a degree of corporate feeling in the CID, you really operate as an individual or in a group of two or three. As a detective you can isolate yourself from those broader issues of what is policing. It is a purely professional activity: there is a crime that is yours to investigate. There are a number of clues, and off you go, and that can be like doing a crossword puzzle. When you are 25 years old you have no great wish to change the role, and I was quite happy doing just that.

Our team was doing a good job. We had got involved in the heavy robbery scene with numerous informers telling us about what was happening, which entailed exceptionally long hours in quite dangerous operations. In the CID there was a view of policing that is to do with crimes being committed and what you do about those crimes. My eyes had not been opened to the importance and implications of policing in terms of the quality of life for all of us as citizens.

As a detective, you actually do not have to worry about the sorts of issues that surround community and public tranquillity. The things that you are dealing with are not great issues in the public domain. If you stab someone in the street and I as a detective track you down and catch you, people are usually happy about that. If, however, in a particular street there is a family who have noisy parties and half of the street go to the party and the other half are upset about it, and the police are brought in, there are all sorts of issues that arise about what the police may or may not do with respect to that party. These issues are far from being clear-cut, and about them there is very little consensus.

If X or Y was behaving in a way which I would not have supported (I am not talking about any dreadful things they were doing, but their attitude to the service), I felt that was a matter for them—let me get on with what I am doing, even though I was uncomfortable with their values and the way that they were doing the work. I am certainly (to some extent I have always been) something of a maverick, and I could not involve myself closely with things. I am conscious that I do not want to go with the herd. If I sit down and think about it, and feel that what the herd is doing is wrong, then I will stand up and say they are wrong and I will go my own sweet way. I have always been like that, and have been like that on occasions in the police service.

D5, a prison governor:

In my early years I adopted a fairly pragmatic approach. In terms of the establishment as a whole, there was a tight hierarchical structure, and perhaps because of my previous disciplined training I recognized the areas for which I had responsibility and which I could change. I also recognized the wider areas, for which I had no direct responsibility and which I could not change at that stage. I was not conscious at the time of being part of a management team which developed policy. I was responsible for part of the prison and was aware that I was not expected to have any wider influence. For the first number of years, I simply got on with things and did not express too many opinions. I did not have preconceptions, but formed opinions quietly over a number of years, sharing views with one or two colleagues who were also growing up in the service.[13] I was never conscious of any feature in the prison that was totally unacceptable to me, although one or two came pretty near it on a few occasions.

The experiences of the Practitioners suggest that there is no predictable pattern to the early phase of their professional career in terms of the development of a working ideology. The one common thread appears to be an appreciation of the need to define the relationship between self-concept and the agency. The retention of this somewhat distanced relationship between practitioner and agency appears to encourage and enhance the development of a Credo Three identity. The next chapter explores further aspects of this process, with particular reference to distinct turning-points in the course of criminal justice careers.

# 4 Turning-Points

THIS chapter explores events that are regarded by Practitioners as being critical turning-points in the development of their working ideologies. Markers may be set down by Practitioners with respect to values and beliefs, and the narratives reveal the variety of watersheds occurring over the course of professional lives.

## Setting a Marker

Several Practitioners identified a particular incident which enabled them to impress a personal value position upon their working environment. Most of these incidents occurred early in their careers, and became benchmarks both in terms of how they saw themselves within the agency and how they were regarded by others.

For E1, a justices' clerk, an incident raised issues that went to the core of the court's work as he saw it:

> Early in my career there was a case involving a defendant who had been convicted of driving while disqualified. I have no doubt that there was a feeling, in parts of the organization, about this particular individual. To my mind there was a question-mark about whether the case would proceed in a fair way. Behind the scenes, people had talked themselves into something of a witch hunt, the sort of thing which can start from small things like tone of voice, then random comments, until the word is that someone is 'a right tow-rag'. The atmosphere was thick with it so far as the police were concerned (this was before the days of crown prosecutors), and the signals were obvious and were spilling on to others in the system. It was something I had sensed on other occasions. The risk is that coded messages and signals start to short-circuit the judicial process. The defendant was kicking up a stink about a procedural irregularity, and this, if anything, was making the situation worse. The problem was that he was right. I was convinced

that the case should be reopened, and that is what eventually happened. The evidence did not satisfy the magistrates and he was acquitted.

Two points are relevant: someone had to be prepared first to lean over backwards to get the original mistake acknowledged, and then to keep the prevailing atmosphere at bay. Perhaps I did not make myself very popular with some of the people involved, but it set a marker which was the important thing. Similar early experiences led me to react if ever messages and signals began to flash, to prevent the system from starting to bend, from whatever cause. That was important and significant, and is a standard that I have always tried to maintain.

B3, a chief crown prosecutor, risked alienating the police by opposing the prosecution of some young people:

Before the inception of the Crown Prosecution Service, when I was working in a solicitor–client relationship with the police, I can remember only one occasion where the police did not readily accept my advice. The case concerned a male defendant in his forties who was organizing a ring of child-shoplifters during school holidays, and the file was passed to me to prosecute the adult and the juveniles. The police investigation brought the whole thing into the open. The parents realized that they were not supervising their children properly. To take the children before the court would not have achieved anything. I gave this advice to a police superintendent during the absence of his superior on holiday and he said: 'No. We shall carry on and prosecute them all.' I lost a lot of sleep over that. I really was upset, and I felt it was fundamentally wrong in those circumstances to bring the children before the court. Therefore I went over his head when the chief superintendent returned from holiday. By then I had been working with the police for about eighteen months, and they were beginning to react positively to my presence, so that, with a great deal of courage, the chief superintendent decided to support me, and the summonses against the juveniles were withdrawn. Nowadays, under the Code for Crown Prosecutors, those children (who had no previous convictions) would have been cautioned almost automatically by the police, and no one would normally dream of prosecuting them.

A2, a chief constable, took a stand that left him uncomfortably exposed when, as a young constable, he distanced himself from a corrupt relationship between police officers and market traders:

I began to feel slightly different. There had been some fairly routine corruption at a low level. It amounted to taking cash from market traders in return for facilitating parking around Covent Garden, which was then a thriving market. I am not sure what offended me most about that, whether it was the fact of petty corruption or the loss of dignity to the service where it was assumed that I would be involved in it. We would be treated with very scant regard by market staff, simply because we were all believed to be in it. It was very small-beer stuff. I was no Bow Street Serpico,[1] but it was enough for me. Not terribly sensitively, I shouted the odds about this, casting myself in the vanguard of an anti-corruption lobby. The chief superintendent gave me the time and the facilities to prepare a report on contemporary difficulties, such as recruiting and wastage from a 'ground-level' perspective. Among the peer group, the report was said to contain the names of those 'on the take'. It did not. In reality, I had done nothing more than express my concern to a superintendent of the day. I was then isolated for considerable periods of time from a widening group which found that the ripples were unacceptable. For example, I would be doing point duty for eight hours at a time and needing a 'blow' [relief by a colleague to provide a pause to go to the lavatory or to take a cup of tea], and I would simply not get a blow. The boycott made me more determined, and that was the beginning of my swimming against the stream, being prepared to stand up and speak for what I saw as some of the wrongs of the day.

The isolation was broken by a constable coming up and offering me a blow, and my saying to him: 'Do you know what you are doing?' And he said: 'I know precisely what I am doing. I support your stance.' 'How many more are there of you?' I said. He said: 'Well a number.' We started from there. It did not solve the corruption—that coincided with the moving away of the market at a later stage—but it established a *modus vivendi* between two different aspects of the canteen culture.

A3, a chief constable, sharply defined the limits of his powers in the face of a national utility company:

When the Central Electricity Generating Board started to move in to test for a nuclear power station, the local people were not amused. All the neighbourhood came and sat in the field so the Board could not get on with its work. The CEGB were annoyed about this and that was understandable, but it is one thing to be annoyed and another to ask the police to turn these people out. I argued that it was not our duty to do that, and that these people were not trespassing. I could have had them all moved and it would have been quite simple. If anybody had objected I could have taken them to court and had them bound over. There would have been no argument, and not an eyebrow would have been raised in the country. The police would have done a good job, the CEGB would have been delighted, and I could have gone fishing. But that is not what policing is about. In any case, the CEGB, who sought an order of mandamus in the High Court, lost, so I was vindicated.

## Watersheds

For some Practitioners, particular events or experiences changed the course of their professional careers. Not unusually, the turning-point arose while undertaking further education, often during the course of a research project.[2]

E2, a justices' clerk, on an early and traumatic experience in court:

A girlfriend came to watch and tore me apart afterwards for being supercilious and talking down to people. I had called a defendant by his surname. She told me, in no uncertain terms, not to be so pompous. That had quite an impression on me, as I did not have an attitude of one-upmanship on the rest of mankind. It sowed a seed I have kept ever since. That was an early shaker, thinking I had an important job in the public eye and doing it well, and minutes later being shot down by somebody whose attention I relished.

An incident of fifteen years earlier stood out as a matter of deep regret in the mind of B1, a chief crown prosecutor:

It was a minor [driving without] 'due care' case where the local chief inspector had authorized proceeding, and I looked at the file and did not think it was on. I had a long and extremely amicable discussion with him, but in the end I allowed myself to be persuaded that I ought to let the case run. Inevitably the case was dismissed, and I thought afterwards that I should not have allowed myself to give in. I am lucky if I can say that that sticks out in my memory after fifteen years, but I always castigate myself for having given in and allowed the case to run when it was totally against my better judgement. This is a very personal thing, but it sticks out far more in my mind than the numerous occasions when we have had heated discussions with police officers across the table who say, 'why the hell won't you prosecute', and we say 'because we don't think it's right'. It is that one instance of a far less serious case, in the early days of prosecuting, that sticks out in my mind, and I should have been more resolute about it.

Particular incidents sometimes stand out as a matter of regret if not of personal shame. D2 had been in the prison service for fourteen years when, as an assistant prison governor, he encountered a darker side of himself:

At the Staff College I had read quite a lot of concentration camp literature, and also had read Zimbardo and Milgram on the potential we all have for cruelty.[3] At a local prison, during a time of peak overcrowding when we went up to 1,300, I found there was a bit of that in me, and that was frightening. The standard occupation of a cell on the wing was one double bunk and a single bed. I talked with a senior officer about the prospect of putting a mattress on the floor, in between the single bed and the bunk bed. We did not have to do it, but afterwards, reflecting on this, I was horrified. I had actually thought that it might be possible to get more prisoners in if we did that. That was really quite frightening.

For his first few years in the police A1 was able to conceal his independent frame of mind. The turning-point for him came during his mid-twenties, curiously while he was a Special Branch officer:

I tend to be of independent mind and had had quite a number of heated, animated discussions with my parents. In the early stages I was very much a police product, a clone of the system.

From the age of 19 through to probably my late twenties, I tended to lack discretion and to be very much the organizational man. I recognized what the rules were in order to advance. I saw things in very black-and-white terms and I responded to the stimulus that the organization placed upon me and which therefore would bring the rewards. I was a product of the system, and I did not stop and reflect. I did not question those values that the organization inculcated in me; I just got on and did the job. As a result, I was recognized and I advanced. It started to change when, in my twenties, I specialized beyond CID into Special Branch, and that led me to new work, regarding all sorts of activities, including trade disputes. Being with the workers at strike meetings gave me an insight into the other man's perspective. The black-and-white perspective that had been induced by the police process was suddenly challenged: 'Hello, this doesn't seem quite right.'

These issues took on a sharper focus a few years later:

As a community relations officer, I was neither a policeman nor a member of the community. I was a bridge, and you have to be a little bit circumspect and questioning, and not only about what the public are telling you. I found that it was not all black and white, but a lot of grey. Whilst in the community relations job for about twelve months, I met a chap from a university. He came to talk to us about community relations and was applauding, perhaps in a patronizing way, efforts that were being made in that force to bridge the gap between police and the community. He spoke at some length about the appropriateness of community relations as a specialization. I had given a lot of thought to this and said it was totally inappropriate: if you really wanted to get to the heart of community relations, then it had to be about the one-to-one contacts on a daily basis between professional policemen and members of the community. One cannot abrogate that day-to-day responsibility to a specialist group. If you want proper community relations, then you have to get to the bobby on the beat.

From this initial contact grew a number of discussions and an invitation to undertake a research degree. I went to the university and undertook three months' intensive training in research techniques and considerable reading for six months. Then back into the force but on a detached basis. (They allowed me a year off.) I camped myself into a non-operational police station and started my research. That was probably the most formative period, out with a tape recorder, talking and listening to people. Although it had become increasingly common and acceptable for officers to obtain first degrees, it was highly unusual at that time to be allowed to research the police organization itself. I had a tremendous dilemma, because what I was discovering was not in accord with what the force necessarily wanted to hear. We all go through the painful process of deciding whether to publish and be damned. I grasped the nettle and decided to publish, even if it led to me becoming *persona non grata* with my chief constable, which it did. People do not necessarily like people blowing the gaff. I touched upon the very sensitive subject of how the organization functions and its consequences. Nobody ever feels comfortable about that kind of exposure, especially practice that contradicts the theory.

I applied to undertake the degree over a four-year period in my own time and I was told: 'You can have a year off and do the rest in your own time.' I readily accepted the offer, but whilst I was detached for my year a new chief constable took over, who was the traditional, metropolitan, seasoned, 'lock 'em up and get results' detective. I went away under one patronage and came back under another, who found what I had to say to be most unpalatable. In fact, he would probably say I was disloyal for exposing the 'secrets' of the organization.

I am quite convinced that if I was still there I would still be an inspector and probably would have been given the run-around for the past fifteen years. I know from people who were in that chief constable's staff office that whenever my name came up his eyes bulged and he went red in the face.

Details of his thesis might later have placed A1 in severe difficulty had he not by then moved to another force:

After the thesis was completed and accepted, relations really went sour. There was a death in the police cells in extraordinary circumstances and an investigation was mounted. Later there was a half-page newspaper feature on the incident. In that article the author referred to research describing how the police forces sometimes use unpopular areas as punishment postings.

The columnist had obtained a copy of my thesis out of the college library and used part of it which described the process. I got a call telling me about the article from a friend, who was another maverick police officer—there was almost a club forming [of those] who had similar views and wanted to bring about change.[4]

Although the thesis was not finished, I was being invited to talk to academic groups. I made an application to my new chief constable for permission, saying that I would be going in my own time and at my own expense. I was regularly refused permission. That was the turning-point for me. I decided that, unless I could move to another police force where the climate was more accepting, I would have to leave the police service.

All sorts of pressures were applied to make me conform. I resisted these and, surprisingly, I survived. I thought through the dilemma of whether I should leave the police service or stay. Was it better to abandon ship and work from outside or to stay and bring about change? My tutor at the university and I spent many hours together talking about what I perceived to be the hopelessness of the situation, with him saying: 'Just hang on in there, keep those values, keep plugging away.'

I started to look for jobs outside the police in social services and in academia. As it turned out, a job came up in another police force. I had been interviewed for two previous chief inspector posts and when the subject of my thesis came up I said what I had to say which led to a row during the interview, and I was rejected. It was just the same in the third interview, but I stuck to my guns and made my point of view. They gave the job to another applicant; but afterwards the chief constable called me back and said, 'I am thinking of creating a job on my staff: how do you feel about it?' He wanted to do something about relationships between police and youth, but he was not sure what. He looked to me as somebody who he could bounce ideas off, and perhaps someone who could present him with new ideas and had a less than traditional approach to organizational and policing problems. He latched on to that part of my philosophy that I was able to bring over at the interview, which was about an alternative approach to policing, not just geared to pure enforcement and numbers of arrests. It was the first his assistant chief constable and the chief superintendent of the division, where the job was, had heard of the post. I am sure he created it there and then. That was the turning-point, because if he had not offered me that job, and thus allowed me to work through my philosophy to my own satisfaction, I am sure I would have been out of the police service.

Personal stocktaking may also be prompted by in-service training courses. A5, a senior police officer, recalls a period at Bramshill Police College:

I was promoted to detective inspector and was dealing with people involved in more serious crimes, sometimes being in charge of an investigation. Then, bingo, I went on the inspectors' course and the scales were taken from my eyes. The course dealt with broader philosophical issues and the policing world in general. It was the first time that I had time to sit back and view policing from other than the internal professional technical perspective. One of the courses I did was looking at policing and control of the police force in various countries, and that opened my mind to the fact that there were other ways of doing it. My six months at the Staff College gave me time to sit back and think, 'Hang on, what the hell is going on? What is the police force about? Let's take it slowly—what are we seeking to achieve?' Life is not just about investigating this crime today and that crime tomorrow, but it is also about making decisions what to do and monitoring them. But according to what standards and criteria?

For A3, a chief constable, Bramshill introduced him to a wider world beyond the police:

My first attendance at the Police Staff College was a real watershed, because for the first time I had access to a good library, time to study, opportunity to prepare and write papers, to present them, and to express my opinions.

I became influenced by philosophers and by humanists. Rationalism, the pursuit of goals through reason, and the use of experience to solve human problems rather than falling back on power exercised by the mind. I have a predisposition for wanting to learn, which I would never have got if I had pursued the police culture and ethos to the top.

A2, a chief constable, describes an inspectors' course at Bramshill and his later reaction to it:

> The atmosphere of the staff college is now encouraging and 'enabling' as opposed to the repressive stultifying atmosphere that existed when I was there. The way that I dealt with the course was by sycophancy rather than rebellion, which was what some people on the course chose. I chose sycophancy and I came away feeling that I had sold my soul for a pass certificate. I have been trying ever since to make good that feeling. It has given added life to my rebellion at odd times, and has added strength to my present insistence on creating an atmosphere for each command structure within which officers feel free to disagree—strongly, if necessary—with senior officers, with our policies and styles. There were officers who spoke out against what we saw as the arbitrariness of the system, and who were failed. Good solid CID officers who were failed by a mark or so on their traffic paper. It felt alarming. I toed the line, and was excessively cautious about disagreeing with those in absolute authority over our lives and careers. In the written appraisal of me at the end of the course there was a reference to over-cautiousness, and it might even have been to sycophancy; but, typically of the system and its lack of forthrightness in those days, I did not discover that for many years after. I had been a prize-winner, and the negative was suppressed. I did not get it as feedback as we left the College, but then I did not need it, because I knew how I had prostituted some of my values.

On being posted as a sergeant to Notting Hill, his immediate concerns were with issues to do with personal confidence rather than questions about the nature of policing in the mid-1960s:

> People were not much asking questions in the way that a sergeant would nowadays. It must have been burgeoning in me, because I called in at the Citizens' Advice Bureau and I recall them saying to me: 'This is the first time we have ever had a sergeant in asking why we are here and what we do.' The Citizens' Advice Bureau had always been a rather mechanical reference point. If police could not handle a problem, we would often refer a third party to them. I had called in to gain a better

insight, but that was a minuscule development alongside all the thinking that goes into police community sensitivities nowadays. It was much more of a football match mentality in those days. The popular police view then was that a local restaurant was being used for drug-taking. The police would raid the Mangrove and get nothing from it. The feeling would be: 'Mangrove 1, Police 0'. It seemed that we would then have to do it again later in order to equal the score, and the score would be equalled by an arrest or two. In the six months that I spent at Notting Hill I was very much in the swim with that. I might have sensed intuitively that it was not right, or product- ive, but I did not have the ammunition to be able to deal with it in any strategic sense. I did not go to the chief superintendent and say, 'This is unproductive; we should be thinking about a new way of tackling policing in Notting Hill.' While there were some signs of a growing sensitivity, I was still very much swept along by quite a strong culture.

A major influence came when it was decided that the Metropolitan Police would introduce something called 'social studies'. The thought was that we had been teaching the 'what' and the 'how' of policing too long without teaching the 'why'. Some people taught the 'social studies' earnestly and with enthusiasm; others simply wrote a suitably philosophical title for a lesson on the blackboard and got on with what they would call 'Real Police Work'. I suddenly felt shrouds being torn away from my visions of policing. That was a major growth point for me. If I could mark myself today as being in any sense different from myself when I joined, the change would have been in that fortnight. I started to see things differently and perhaps more liberally than I had seen them before.

C2, a chief probation officer, on turning down his first offer of promotion:

After seven years' working in my first post, I wanted to explore and develop skills in community development, and so I asked my chief officer whether I could have a year's secondment from the service. With some reluctance, but with his usual support, he agreed. I went to the National Institute of Social Work and worked on a community project in a homeless families block for about ten months for two days a week. That was a very important year out. The interesting thing for me during that period was being dispossessed. I had enjoyed the status in a role of being a probation officer and was free to go into homes, hospitals, schools, Borstals, and so forth wearing that title. Now I was just a detached worker, working for the Southwark Community Project in a pioneer model for community development projects, and my only authority was my skill and ability to make relationships with people who felt powerless and disadvantaged. You were only as good as your last call at somebody's door. That was a very important lesson, and I feel privileged to have had that experience because somebody described me as carrying my authority lightly in the service, and part of that goes back to the experiences of working on those estates.

I felt profound misgivings at the end of that year. I had been to lots of RAP and NACRO meetings and was in touch with all those circles.[5] That was now part of me. In many ways I came back as a stranger to the probation service. I wanted a leadership position in order to bring some of this sense of innovation and excitement about discovering the community, both in the neighbourhood and in the inter-organizational context. I wanted to get that into probation thinking itself. I also had ideas about an experimental day centre and how it actually could be run. But it was not to be. I felt harshly treated because I was given a case load of sixty on an estate in Nottingham and told to get on with it.

In less than two years, however, C2 had been promoted and was given the task of setting up one of the first community service order schemes, established under the Criminal Justice Act 1972:

I used the word 'community' [in the community service order] in a way that others were not using it. The Wootton Report talked about the probation and community service orders as being both punitive and reparative, but also talked about building bridges with local communities, which is the term that I hung on to, and I took it as a reconciliatory gesture.[6] My chief officer sometimes accused me of being idealistic and not having my feet on the ground, but on the other hand he knew that I could generate the ideas from which could come some working reality. You have got to make some compromise in order to work these things. I left after nearly two years, but it had been an intensive and exhausting experience. It needed a consolidating organizational thrust, and the maintenance of the organization became as important as some of the innovatory aspects that I had pioneered. Some people are innovators, while others are maintenance people.

C3, another chief probation officer, also took the risk of being isolated by working beyond the boundaries of the agency:

> My experience over four years as an assistant chief probation officer, given my orientation to social work and public service, was almost ideal, because working across agency boundaries I had a role to develop facilities for offenders. At that time I did a lot of work with some NACRO staff and have been actively involved with NACRO ever since. It was about deliberately working with voluntary sector organizations and with the prison service. I had specific responsibilities in relation to Crown Court work, community service which was a very rapidly developing part of the service's work, and innovative work on through care. I was a bit isolated at times because I was out there, working across boundaries, and did not know whether I could carry the service with me, and had to be a bit careful about that. But it gave me the opportunity to see the potential that there was for a much broader constituency of concern about offenders and what might be done, rather than how it is sometimes seen from within the probation service. The probation service suffers from some of its narrowness of outlook.

## Recognizing the Damage of Custody

For several Practitioners an awareness of the limitations of criminal justice was especially sharpened by direct contact with prisons. Decisions were taken in the 1960s and early 1970s that led to about 500 probation officers working in prisons and young offender institutions. The 'welfare' role of the probation officer in prison meant engaging directly not only with prisoners but also with prison officers. Some probation officers became closely identified with the prison management, while others were careful to maintain their distance. Whatever the merits of probation officers working within prisons, by the 1980s large numbers of them had a close appreciation of the prison system. For some probation officers, this experience contributed to a profound belief in the failure of the prison. For other practitioners working across the criminal justice process, contacts with prisons, although often sporadic and brief, have become more frequent. A result of these opportunities to get a fuller view of many aspects of the prison is that the ideology of prison systems has become increasingly challenged. This exposure, as Thomas Mathiesen has suggested, breaks down the pretence that hides the 'fiasco' of imprisonment. Mathiesen's argument is that 'we have prisons, despite the fiasco, because there exists a pervasive and persistent ideology of prison in our society'.[7] This ideology, which renders the institution of the prison as meaningful and legitimate, is supported by 'an inner circle of criminal justice personnel, who pretend that the prison is a success, though in fact they more or less know that it is not . . . [w]ithout it, much of the work done by people and institutions within this sphere would appear meaningless and counter-productive'.[8] In Britain, during the 1980s the inner circle of élite practitioners was penetrated by persons who were willing to abandon this pretence.[9]

C5, a chief probation officer, worked in a young offender institution at a fairly early stage of her career:

> It is much better to be in a position of having experienced what it is like to work in an institution, particularly if you are challenging what institutions are all about and what they stand for. One can then speak with some authority which helps. I really do hate institutions, but I did not hate them until then. As a field officer, I used to quite enjoy visiting institutions as part of the job, but working there I realized it was so repressive, inhuman, dirty, and noisy. It made me realize why I could not cope with being locked up. It helped me to understand how people do suffer and why they react in the way they do in institutions. One of the things that appalled me most about the place I worked in was how many people just accepted it. I felt it was more healthy for those who had actually kicked against the system.

As a solicitor, E4's introduction to prisons was especially sharp:

My knowledge about prison was negligible. Like many of my practising brethren, my interest in the matter stopped at the prison door. By the time the guy got sentenced to imprisonment my legal aid certificate said I could simply advise on whether or not he had an appeal. I would go through the routine and send him a letter in prison. I did not even have a clue where he was. I would go into prisons periodically to interview people who were remanded in custody, but that was really it. Then somebody poured boiling water all over a client and I wrote to the Home Office: 'Dear Sirs, a terrible thing has happened: my client has had boiling water poured all over him. I obviously would like to know more about this and I am instructed with connection to the claim because it is said that this happened as a result of a man getting hold of equipment to do something that prison authorities were there to prevent.' I got this extra-ordinary letter back which said: 'What goes on in the prisons is of absolutely no concern of yours, and what is more, if you attempt to discuss it with your client we will stick a prison officer in with you to listen to your conversation.' I wrote back and said: 'I think you have misunderstood the situation: I am instructed to make a claim.' They wrote back: 'Please notify us in writing of your intention to attend the prison on the next occasion.' And sure enough, there was a prison officer there. It was absolutely unbelievable.

In a series of prison cases, he describes his success in forthright terms:

I started thirty-two cases in the European Commission of Human Rights on denial of access to a lawyer on behalf of people who had been convicted of terrorist offences. For ten years I had great fun running rags round the Home Office, and eventually broke the back of the simultaneous ventilation rule and the pre-ventilation rule that they had invented. I learned then that the system is built one way, as a wall against the mass. You breach that wall and very often there is no inner wall, and they have to construct it. If you turn the system round and play the system against itself, it has no defence because it has never looked that way.

B3, a chief crown prosecutor:

I believe prison is the last resort. It is very negative in someone's life if all we can do is lock him up and physically prevent him committing offences against the community. One should seek to reform if one possibly can, and reform rarely takes place in prison. In the very rare case a man can be so full of remorse that he decides he is never going to offend again. But those cases are, in my experience, few and far between, and there seems to be statistical support that, the more people go to prison, the more they become entrenched in their view of themselves as criminals so that when they come out they carry on committing offences. I would rather see initiatives such as community service for keeping people out of prison, but at the same time criminals must be made to realize that society cannot let them continue to act in an anti-social way. The danger is that it will go too far, and thus we get swings and cycles in social history.

C1, a chief probation officer, relates his experience of working within a prison to other interventions by the criminal justice process:

Two things changed during my second job as a probation officer. There was a much more sober appreciation of what prison does to people and a recognition that we got wrong a lot of what we were doing in prisons. The day centre, with which I was very closely associated for quite a few years, also made me stop and think long and hard about youngsters who had been institutionalized from an early age. My ideas on juvenile justice shifted a lot because the youngsters who had come out of care were totally unable to manage and their experiences in approved schools had been outrageous. Dealing with the end-products in the day centre started me thinking that we had got that wrong too.

This chapter has highlighted episodes occurring during the course of professional careers that were important both in terms of defining positions on issues and in shaping the accommodation reached with the employing agency and the broader enterprise of criminal justice. The next chapter explores further aspects of the relationship between practitioners and their employing agencies.

# 5 Individuals and Agencies

THE development and consequences of criminal justice practitioners' working credos cannot be adequately understood in isolation from the employing organization. But not all practitioners are employees of public agencies. Solicitors and barristers, for example, often work with a handful of colleagues and in some instances might be regarded as self-employed. The involvement of lay magistrates in the courts is part-time and unpaid. For others, important differences exist across the employing bureaucracies. As civil servants, prison personnel and members of the Crown Prosecution Service have to work within the constraints emanating from ministerial responsibility. These include restrictions on expressions of personal views and public statements about their work. The situation for probation officers, court personnel, and members of the police service is less clearly defined. Some latitude arises from being employed by locally administered agencies that are largely financed by central government, and this allows valuable room for manœuvre for the reform-minded practitioner.

These variations as to form and structure of employment need to be borne in mind in this chapter's exploration of the sometimes tenuous and fragile fit between the agency and the Credo Three practitioner. There may be aspects of existing arrangements from which the practitioner wishes to be distanced, as well as basic differences regarding objectives. As a reformer, the Credo Three practitioner often appears to be swimming against the tide. Seeking and sustaining change may be the source of frustration and disappointment. Indeed, efforts at reform may sometimes give way to a struggle for sheer survival. The perennial challenge for the Credo Three practitioner is one of avoiding co-option, marginalization, or rejection by the organization. The Practitioners, all survivors in their professions, had reached an accommodation with their agency that enabled them to continue to their own satisfaction. They had progressed to senior positions and, for the most part, did not envisage a career change before retiring. Dedication and competence made them valued employees, and, in some cases, their personal adherence to Credo Three and the creative tension thereby generated was appreciated by the agency. This outcome, however, should not conceal the typically uncomfortable and ambivalent relationship between the Credo Three practitioner and the employing agency.

The general problem is cogently set forth by A1, a senior police officer:

> If there is something that underpins my personal philosophy, it has been how the organization influences the behaviour of the individual. People can become almost automatons as they follow the organization's line.
>
> I have spent a lot of time thinking about the incompatibility between the needs of the individual and the organization; much of my work is about their mutual adjustment. The organization tends not to be very adaptable and it is therefore the individual who has to adjust, and that can and often does lead to alienation. My research found a high degree of alienation in the police because of the difficulty that individuals had in equating with organizational ideals and objectives. Doing the research was a major turning-point for me and for how I worked. Before doing the degree, a report at the Police College described me as reticent. Having gone through the painful process I have described, I became much more confident, to the extent that I have probably become, in police terms, a 'maverick'. The training gave me the confidence to voice my opinions and, to a certain extent, to be constructively critical within the organization. Mavericks are not well received in the police service.

When D2 became an assistant governor in the early 1970s, some people regarded his appearance as somewhat unconventional:

> I did not see myself in the mainstream of young assistant governors. If colleagues were to have drawn a cartoon of me then it would have been of an academic hippy, with hair down to my shoulders, wearing sandals, a beach shirt, and beads. It was never as extreme a picture as that. In those days long hair was quite common outside the service. I did wear a beach shirt

from time to time, but the Borstal was on the east coast, and in the days before staff were in uniform it was quite common in the summer for staff not to wear a tie, and to wear sandals. But the folk memory is that I was the eccentric and the rest of the staff were in three-piece suits.

At the first Board of Visitors' meeting I attended the chairman, a really wise man, asked to see me after the meeting. He asked me why I was wearing my hair so long and I said something like: 'Well most of my life is spent outside the Borstal.' He wished me well and said that he was sure that I would settle in. Then he said: 'Remember, it will be harder for you to achieve things than it will be for others because you have to overcome this barrier of how people see you.' That was something that I had to work at. In time, staff knew that they could trust me. Once they saw that I was professionally competent, the way I presented myself took a bit of a back seat.

Twenty years later, he was looking beyond the agency to legitimize his role as a prison governor:

At present I have not agreed my job description with my line manager. I am niggling over a sentence which says something like: 'You will manage the institution according to those departmental regulations in force from time to time.' I am arguing that I am quite happy to say that I will manage it according to the law, but not simply in accordance with departmental regulations. So many times in the past, departmental regulations have been found to be contrary to the law. I owe (it sounds very noble) a higher duty outside the Home Office to get it right.

Early in his career as a probation officer, C2 was regarded as a risk-taker:

My first chief officer once referred to me as a butterfly, a risk-taker, and somebody who always operated on the boundaries of the service. But he had sufficient confidence and admiration for what I delivered on the bread and butter. As a report was never late and as my duties were conscientiously done, he was prepared to see me operate on the margins. He identified with all that because he took risks himself.

Some years later, C2 sought explicitly to offset his maverick instincts, and to acquire the political skills of compromise and concealment:

> I deliberately took on the staff job as deputy chief probation officer because I had experience in innovation and community development, and was regarded as something of a marginal flier. A good organization cannot run on mavericks, and I needed solid experience in terms of running the staffing side of a large organization with all its complexity. It was what I least wanted to do, but for my career I actually had to learn how to run a large-scale organization which employed over 400 probation officers and had a total strength of just over 1,000. I recognized that I had maverick instincts and was not in the traditional sense an organization man. I needed the experience of coping with grievance codes, employment, and redeployment issues, and of mastering industrial legislation.
>
> I learnt political skills in terms of handling large committees, relationships with county councils, the intricacies of funding, the overlap between areas, and how things could easily go wrong if you have missed some of the steps along the way in terms of developing policy. I became far more sophisticated about political realities, the chicaneries, the compromises, the deals that were done behind the scenes, not being naïve, and going into it with open eyes. I became something of a political opportunist as I did not have those instincts before.

At mid-career point, A2 had a strong personal conflict as a senior police officer who became caught up in the Notting Hill riot of 1976:

> The racial context was very clear to me. That pressure was very close, because it culminated in the Notting Hill riots where I felt particularly uncomfortable. I had been a community liaison officer and I saw my role as preventing riots. I had failed. My advice was that we should consider very seriously what sort of policing profile we adopted for the annual street carnival. I was quite happy to be robust about the criminals that were involved, but I felt that a generalized heaviness of policing was something that we should avoid if we possibly could.
>
> It was not a popular or acceptable line of argument for the

day. A riot ensued, and I still believe that the 'profile' of our policing was one of the precipitating factors. It was all the more difficult because I was given an operational role as a super-intendent to play on the streets during the riot, so I felt a tremendous split between my community liaison self and my operational self. I was able to discharge the two roles only with some difficulty. The adrenalin ran high, and it was not a very dignified time for the police, either in the sense that we had to grab dustbin lids to protect ourselves, or in the way that we handled the riot as a police operation. Most riots are chaotic and messy, and public expectations of police ability to quell them are unreasonably high. I was aware of a couple of col-leagues who reacted very well professionally and who seemed to be quite calm and contained. I felt that I was not much like them. It was a moment of profound regret and disappointment at myself and at the profession.

But the communities of Notting Hill rallied round in a very supportive and positive way and were determined, with the police, that a riot should not happen again at that level of severity. They have succeeded over the years. The police com-mand has changed enormously and is much more enlightened now than was my perception of it then. My failure was not to have networked sufficiently well for there to be clear sanctions in advance on rioters from responsible members of the com-munity, and my inability to persuade the police command of the day that a blanket approach to policing was inappropriate.

In his pursuit of the ideals of justice in defending persons charged with terrorist offences, E4 has been at pains to respect the formal conventions of the court. This stance, however, has not prevented him becoming very isolated within his own profession:

I have never employed what might be loosely termed 'left-wing' counsel, largely because their view of the function is incompatible with mine. In taking instructions I may have to say to the client that there are certain things that I will not do. In terrorist cases I will not allow myself to be used as some kind of a political flag-waving exercise. I say: 'If you want to come to court, and are prepared to treat the court with respect and be tried according to law, you will get the fairest trial I can get for

you, but I'm not going to stand there and say "Up the IRA".
You can stand up and do the clenched fist salute and say "Up
the IRA" as many times as you want, but don't expect me to do
it.' I have always found great support from the counsel I have
instructed, who have been intelligent, caring, and competent,
and some of whom, as a result, have suffered in their careers.

He frankly spells out how he was isolated within the legal profession and was, he believed, branded as a terrorist by the security
services:

You can become very isolated if you take on a cause of this
nature. My professional brethren were never overtly hostile to
me, but there came a point where they felt I had done enough
and that I was being fanatical to continue. That was where
we started to part company on what was important so far
as principles were concerned. I learnt paranoia and what it
is like to be followed, to have my office burgled, and to be
under surveillance. I have an MI5 bug file and a profile on the
Lisburn computer. I have been branded as having espoused the
Republican cause simply because I was prepared to do what
I was professionally obliged to do, which was to offer legal
services to individuals who stood charged with an offence. One
is talking about whether one does this kind of work or whether
one does not, because the penalty of doing this kind of work is
that you will be branded with the politics and/or the crime
of the person that you are representing. Pat Finucane is an
example of what happens when it gets out of hand.[1]

I had a long period of time where I had to cope with my own
fear and with the wrath of authority. I am beyond that now. I
have conquered that kind of fear. I do not have any fear of
authority or of police officers. At the end of the day one has to
have a fear in the sense that, if somebody is going to take your
life, you may fear that they will do it. I am not afraid of them
any more; they cannot frighten me. They have done most of
what they could do to me, and at the end of the day have
descended to the level of the people that they are trying to catch
for these offences. If that is what they want to do, then they can
do it, but it does not make them any better for employing those
kind of tactics.

For the reforming prison governor, an especially sensitive problem is pursuing change without alienating large sections of the staff. D4, a prison governor, summarizes the general problem:

Prison governors have become insulated and alienated from the prison community in which they serve. I have in the past experienced a great deal of opposition from staff because I talk with prisoners. It is believed that a prisoner should only talk to the governor after he has made an official request through the proper channels. A governor who talks to prisoners is said to be undermining the system and the authority of staff.

This theme is developed by another prison governor, D3:

My second posting was a time when I was most confronted with my own impotence. For the first time, I was at the centre of the management and had begun to learn a good deal about the importance of management teams. I encountered a POA branch that was intransigent, stubborn, and reluctant to change.[2] The detailing of staff, which I quickly had to learn a lot about, was disgraceful. I found myself at odds with large lumps of the nick, which had been frankly deceiving everybody for some time. Unlike I had felt before, I was the proverbial lone voice, and that is not a comfortable or effective role. It is not good being the only person on your side, even if it is the right side.

And later, becoming the governor-in-charge at another prison, he overtly distanced himself from his predecessor and the existing institutional regime:

The ethos of the place was Victorian. At best it was paternalistic: nice, grateful prisoners were given perks, but nasty, ungrateful prisoners could be left out in the cold for ever and a day. At its worst, it was (I choose my words with care) unhealthy. There was a nastiness about some bits of the jail which had more than tacit approval from the then management team. My predecessor was a very strong and forceful personality, with a view that things are pretty black and white. People have to conform or nasty things happen to them. When I arrived I felt myself very dramatically at odds with that culture, and during the brief overlap with him I took a conscious decision, after the first day, not to accompany him around the prison. I very specifically did not want to be associated with some of what was happening and which I saw him as either condoning or encouraging. That culture needed to be changed, and I wanted to shift it away from the very simplistic and inappropriate notion of just deserts towards a rather subtler appreciation of what people are about, and a recognition (as I try to express it to new staff) that what we are about is giving prisoners more than a fair deal. That encapsulated what I felt needed changing at the prison.

Lots of things happen when you take charge. I was quite daunted at the prospect of being the governor. Part of what I felt at that time was a need to be clearer about some of my own beliefs than I had felt it necessary to be before. In the past (with some exceptions) it had been possible to sail along fairly straightforwardly without having to nail my colours to the mast. In the face of what I saw there (it wasn't all negative: there were lots of things I saw that I thought were positive, such as the sense of confidence which the place exudes and that is a great basis to build on), I felt constrained to go public on some of my views. I am questioning that now, and have been doing so ever since the early days. Part of that learning has derived from the work with the staff association, because one of the things that I get from them is a degree of honesty which one does not always get from immediate management colleagues who like to break things to you more gently. Very early on, the POA were keen to know where I stood; I tended to tell them, and they did not particularly like that. Our paths diverged and have run parallel ever since. The senior probation officer (the nearest I ever had to my own counsellor) made the point to me (and I accept the force of it) that on the grid 'I'm OK, you're OK' ... I have a tendency to drop into the 'I'm OK, you're not OK' box. And, while that is a fine stance for crusaders, it is not a good stance for someone who wants to manage and lead. Perhaps I have recognized that rather too late in the context of being the governor, because I am never going to persuade the staff collectively, and particularly their elected representatives, that we have sufficient common ground to build on, because they see me as an extremist and as a loner. Because I am pro-inmate, the presumption is that I am anti-staff. I do not believe that is true, but it is the perception that is abroad, and I do not believe I will change it in twenty years. It manifests itself in all sorts of ways, but the most striking evidence used against me is that I am soft on adjudications. Where my predecessor would automatically start with seven days' remission, and more likely fourteen, I generally impose a much lower level of awards. That is one touchstone of whether the governor is 'one of us' or not.

The POA vote of no confidence in me happened without a particular lead-up. I had been trying to make some changes in the deployment of staff because, like most jails, there are little pockets of staff being wasted or underused and other places where there is a critical need. And I had been trying to push them to make some changes and they had resisted. Some bright character at a POA branch meeting proposed a vote of no confidence, which lots of people have had at one time or another. It was a quiet news day, and we have a mole who sells stories to the local paper. It snowballed, and I woke up in the morning to hear of my vote of no confidence on the 'Today' programme. It assumed a greater importance than it might otherwise have done and I found that I was having to talk to journalists. It was not a deliberate tactic on my part, but I fell into the line of saying: 'Yes of course it was hurtful to know that my staff had no confidence in me.' That made big headlines in the local press. I have never quite got right whether one should be the forceful, tough character or present the human and vulnerable front. About a year later the secretary took the trouble to let me know that the vote of no confidence had been unanimously reaffirmed.

E1 describes his initial impressions on becoming a justices' clerk:

I was working in an atmosphere where the voice of the magistrates was not really being heard. This was not unusual because there is a strand of clerking which is to do with keeping magistrates in check and of confining their role in the organization. This has receded somewhat, but when I first became a justices' clerk proper I worked with a bench which was accustomed to the idea that, largely speaking, and possibly with the exception of the chairman's special privilege, things started and ended in court. This was not untypical of that era. The immediate task in hand is made easier if magistrates are not lively or active in the wider criminal justice field. The thing they wish to do, the letters they wish to write, the meetings they want to hold to improve matters in their eyes can cause a large amount of work. They certainly did at the start, when I responded by loosening up on the traditional attitude. The long-term effect is that channels of communication, discussions, and campaigns make the situation much more manageable rather than less and in some instances improve things dramatically. Nowadays this is the norm in many courts.

Once in the driving seat, I resolved not to set out to obstruct what people wanted to do but to listen to ideas and concerns; not to say: 'No . . . you cannot do this or that' or that such activity is 'outside the rules'. There is a difference between having a dead situation in which people are simply told 'That is not on' and a live situation where they are at least allowed to voice opinions and discuss matters. Often very good ideas emerge. I preferred an 'open door' approach to one of stifling things. My job was making sure that magistrates remained on the right legal tracks, rather than obstructing what they wished to do. I was not thinking that there is a single idea to be achieved, but freeing things up to see where they went. What I may not have bargained for was the strength of some of the personalities involved. People of immense influence and personal qualities saw this as an opportunity to get things done in the more liberal atmosphere that was created. Messages were coming in which, in the past, may have been ignored or deflected. There were confrontations with others in the system, at the start a large number of these. The rules had changed and not everyone understood this. There was a pressure of ideas and this may have made some officials in other agencies defensive and caused them to wonder what was going on.

An example was the juvenile justice campaign in which E1 was actively involved in the early 1980s, first at the local level and later nationally. He was quick to appreciate that, strategically, among competing reform issues, juvenile justice had much to commend it:

There is a difference between having an idea and getting it off the ground. I think that many people fail at this point—when a project is not driven forward. If you have both an idea and the ability to act on it, and the stamina to stay with it against other people's discouragement and criticism, there is a fair chance that something will materialize. That certainly happened with the campaign about improved services to juvenile courts and resources to support progressive approaches to sentencing. The tag 'juvenile justice' had not been invented, but many people were steaming about the situation in relation to juveniles. My own naïvety about the internal politics of the provision of services for juvenile offenders may have led me to ignore some of the difficulties and hurdles. Whether something can be achieved tends to crystallize in my own mind fairly rapidly, and it then becomes a question of the best tactics or route. Generally speaking, it is necessary to move fairly rapidly, before things go cold or other people have ideas for filing matters away.

In the juvenile sphere there was real scope to achieve something, largely because of the failure of local authorities to come to grips with the Children and Young Persons Act 1969.[3] It took three years of constant pressure to achieve the first indications of success. The juvenile justice issue, locally, got off the ground because magistrates had broken the mould. They had got away from the idea that it had to be left to the regular agencies to initiate everything. They became used to direct liaison with others in the system, to the idea of exerting pressure on occasion. At the time this was considered to be a risky thing to do. It is important to maintain judicial independence, and 'the Lord Chancellor's letter' set out what it was prudent for magistrates to do or not to do in this regard. This was a document kept by justices' clerks in their top drawer, so to speak, ready for use as required. At that time, meetings with social services were actively discouraged on the strength of that letter. In contrast, the more liberal attitude which emerged at my own court, and the basis of initiatives taken there, has had a lot to do with reinterpretation of official instructions, guidance, and sometimes even statutory provisions, to match changing circumstances. It is the difference between using materials in a positive or negative fashion.

The preceding extract draws attention to a variety of issues that arise with respect to court reform. Especially delicate issues continue to arise with respect to sentencing, and much of the resistance to change by persons working within the courts reflects the traditional isolation of the courts from other parts of the criminal justice process and from society generally. Although Parliament began to take a slightly more robust stance with respect to sentencing policy in the 1980s (evidenced by the Criminal Justice Acts of 1982, 1988, and 1991), many members of the judiciary have remained aloof from discussions with other criminal justice personnel. This situation places a special responsibility to make those wider connections upon practitioners who are themselves working within the courts.

E6, a recorder, seizes opportunities to act upon her conviction that sentencing levels are too high:

> I concern myself about whether I am doing sentencing right, and I am not inclined just to look up the guideline cases and act accordingly. David Thomas's *Principles of Sentencing* has been something of a disaster, as far as levels of sentencing are concerned, because Thomas only analyses sentences in the Court of Appeal.[4] People only go to the Court of Appeal if the sentence is clearly over the top, and then the Court of Appeal may knock a bit off, but it is still probably higher than you could very well have got in any of the hundred different courts sitting that same day in other parts of the country. I rather cynically tell new assistant recorders to 'read Thomas if you like, then halve it and knock a bit off'.

The justices' clerk plays the key role in magistrates' courts with respect to sentencing training. E2 approaches this challenge with tact and determination:

> As a trainer of magistrates, I have tremendous influence. From when they are first appointed, I try to make them aware of such factors as the very small number of violent offences. I encourage them to be realistic about what they can achieve and suggest that we achieve very little at the end of the day. I show them the graph of peak ages of offending and say that, generally, if you can contain people between these ages, they are going to grow out of it. That is very useful in the juvenile court and is beginning to have some effect in the adult court, because we are sending fewer people to prison and making more probation and community service orders. One has seen a shift, which will

continue the more you emphasize that custody is special, the more that you go through the hoops of saying, 'Well is this really so serious an offender?' It is getting the idea across that, if custody is the ultimate weapon, then you do not use it until you actually have to.

At a more general level, he decided to push gently after taking over from his predecessor as justices' clerk:

After about a year, both my predecessor and the bench chairman died, and although I had great feelings for them and shed a tear for them, I felt both release and relief that at last the tremendous constraints I had worked under were removed. The only drawback was the need to maintain the confidence of the magistrates and to know from experience that too much change too quickly was likely to alienate. It was very much a process of going softly, preparing the ground, and almost changing them without their realizing that the system had changed. There was so much to change in terms of the administration of cases, how advice was given in court, and with regard to training. In more recent times the magistrates have looked back and seen quite how much has changed. This is rewarding for me, although at the time it seemed very slow and quite frustrating.

E5, a magistrate, learned from his own experience the 'art of pushing slowly' on sentencing, but also with respect to court reform in general:

You can push gently and after five years you may get your ideas over, but if you push too hard you get back to minus rather than just back to zero. I have tried to be fairly subtle in my pushing for what I think to be right. I regret that I took a little while to learn about the art of pushing slowly. I was probably a bit too impetuous when I was younger, particularly in trying to change attitudes on my own bench. When I joined the Council of the Magistrates' Association I had been on the bench for just over ten years, and by that time I had learnt a little, and was a bit more cunning with the Association than I had been with my own bench.

This more cautious tactical approach is also preferred by C3, a chief probation officer:

I have some pioneering zeal, but I am quite good at picking up on that within other people. I like a committed working group around me and I have strengths in enabling other people to make change as well. I do not just see myself as an individual change agent. I like to build on what is going on. I take my time to read what is going on in an area, and therefore might on one or two occasions have disappointed some people who expected me to be a bit more of a new broom who was going to bring in a rapid change. That fits with my casework approach to the client group that I was responsible for. Unless there was some outstanding requirement on me, for example a public protection issue, I would look for what I was actually capable of moving and seek to achieve progress there. That is my management approach too.

Given his view of the police as a mechanistic organization, A1, a senior police officer, sought to encourage formal cautioning as an alternative to prosecution by consulting with and gaining the support of those staff most directly affected. Indeed, introduction of the 'instant caution' required a totally different style of managing change:[5]

I had particular concern for young people as offenders. I was very much concerned about the way in which we caught people up in the system, labelled them quickly, and took a mechanistic and enforcement-orientated view towards them. The outcome did not really matter in terms of whether you diverted youngsters away from crime. What mattered was that you actually captured them.

Managing change in the police organization traditionally tends to be by the rule-book and edicts from the centre. Somebody in headquarters has an idea, works it up, presents it as policy, and it is promulgated. Sometimes there was some consultation, but ten years ago it was the edict approach. As a newcomer to the force, I realized that unless I took people along with me I was not going to get very far. This was because I did not know how the force actually worked and how it would integrate the innovation into the most conservative part of the organization, the CID. I had to find a way to sell the 'instant caution'. The only way forward was to give people ownership of the idea. I said to the youth case officers, one on each division, that perhaps their objectives and mine could be worked out together. The majority of them, especially as they were parents themselves, were attracted to the idea. I had ten disciples who would go out and sell the idea. It was very much an exercise in participation. If you can get people to have ownership of something it is always likely to be much more successful—and, more importantly, to be sustained—than it will be by relying on the edict approach.

I always felt that things stood a better chance if people were truly consulted. It was even better if they designed the product, because being involved fortifies the decision to implement. We created a demand for the system because very quickly the word went out that this is good. One had to recognize that they did not all necessarily fully identify with the philosophy, but could see the pay-offs, such as reduced paper work and a much quicker system. Everybody has objectives. If you can see what these are and meet them at the same time, without clashing with the overall philosophy, and it achieves the greater good, it is well worth doing. We therefore negotiated the entry of the idea into the organization which would otherwise have been resisted if we had imposed it. There is a moral dimension, but I believe the greater good was worth that negotiation. If you wait for the perfect time you will never get there. It has to be a process of evolution and development, but with compromise.

The constables and sergeants saw the practical benefits, and many of them identified with the philosophy. The more senior the officer, the more entrenched the views were to the traditional style of 'hang and flog 'em'. Indeed, some of the senior officers were very, very caustic about the idea of 'letting people off' before the delayed process had created enough pain to sink the message home. The thinking as stated by quite a number of senior officers was that the bureaucratic delay was part of the punishment process. Indeed, the punishment aspect was taken very seriously. For example, we found that in some places police cautions were undertaken in the courtroom adjoining the police station where the cautioning inspector would sit in the magistrate's chair. It was absolutely amazing.[6]

At times it was very lonely in the organization. However, as the group got stronger and started to bond I received support from within. I always recognized that I could get support from the chief constable, but that was not necessarily the best form of power that I wanted to exercise, and in any case it is in total contradiction to my philosophy of involving people. I often say we should turn the police into a workers' co-operative, which is a bit of a 'red rag' to many in the police service because they do not necessarily like to hear that kind of suggestion. The police rely very much on the hierarchical or military model, where basically the sergeant tells the constable what to do. It is not as strong as it was ten years ago, but there is still the feeling that: 'We are not here to consult you lad: you're here to go out and do it.' There are obviously certain policing activities, such as major public disorder on the street, where you do not necessarily want to hold a seminar to discuss the best way forward. There is a conflict within the organization in terms of having, very occasionally, to revert to the military model. The police organization tends to favour the military model rather than the consultation model, and when the chips are down it often reverts to 'orders' to achieve things. Individuals within the organization have certain expectations of you which may not permit you to operate the open style, or if you do they are suspicious of you because it is not typical. That was particularly noticeable on advancing from chief superintendent to assistant chief constable. I could get away with it more as a chief superintendent, but at ACPO rank there is much more distancing. A major organizational problem is communication between the top team—the policy group—and the people who ultimately make it work through middle management.

The complexities of reforming the police are mirrored within the prison system. D5 describes involving uniformed staff and prisoners in shaping departmental policy with respect to long-term top security prisoners:

We were looking to build on the principles that had been laid out in an earlier document which had set out to provide a theoretical basis for good practice. The spotlight turned on the difficult long-term prisoners because they were the ones who posed the greatest threat to the stability of the system. We decided to have a two-day seminar to discuss how best to help these prisoners cope with their sentences. We also recognized that the group of people who had a particular right to be involved in such discussion was the prisoners themselves. In order to make this possible, it was agreed that the seminar should take place in a prison. There were fifty participants: ten academics, ten senior administrators, ten staff of all disciplines from other prisons, ten prisoners, and ten staff from the prison. Locally, we laid aside three weeks of time, for a couple hours each afternoon.

The twenty staff and prisoners met as a group to discuss and prepare the agenda which they wanted to ensure was covered at this conference. In effect, they ended up with a joint agenda. They were on their home patch, and the other people who normally would have been confident were a bit uneasy in that sort of environment. For example, administrators who knew each 'difficult' prisoner as a file that crossed their desks were suddenly faced with the prospect of sitting beside these prisoners for two days. Academics and others who were accustomed to speaking *about* prisoners, and, in their own estimation, on behalf of prisoners, now found that they had to speak *to* prisoners. In the event, the prisoners took over the debate in a fairly articulate and positive way. Some of the thinking of *Opportunity and Responsibility* was formed by discussion over this period.[7]

D1, a prison governor, emphasizes the immediate requirement of establishing a mutual understanding between staff and prisoners as the basis of the regime:

We were bringing the cons into the discussion as to why they behaved the way they did, and trying to encourage a degree of, if not camaraderie, at least group responsibility. It was as close to Radzinowicz (rather than Mountbatten) as I could get.[8] You had to wheel and deal with the prisoners, and you always knew there was a bit of cannabis floating around, but if it was LSD you jumped all over them. Maybe you said, 'OK, a bit of cannabis, and a bit of booze, but no hard stuff', and most of the time they kept to that. There was the normal wheeling and dealing that has to go on. We developed personal relationships with a lot of the cons to the extent that we were probably running the best dispersal regime at that time that had been known. More than anything else, we were personalizing it with many of the men. That takes an awful lot of time, and I believe the governor has to get out and be seen to be concerned and up front all the time. We have got it wrong now. I am convinced the present concept of governor is flawed, as they try to get us to operate as civil service managers which we are not.

The reform strategy may involve going outside the practitioner's own agency, as exemplified with particular reference to sentencing by E1, a justices' clerk:

It is better to go ahead and do something rather than to complain about it not happening. Within reason, I am free to administer the court as I see fit and provided, nowadays (but only with recent effect), that financial and quality targets are met. Justices' clerks also have a free hand so far as their approach to wider criminal justice issues is concerned. I have always given this aspect high priority. Across the country all sorts of schemes and initiatives are being mounted by clerks all the time, but largely speaking, they tend to focus on administrative matters, although not exclusively. But, as a general rule, initiatives by clerks in the wider criminal justice field, as opposed to such matters as management, training, or technology, are less common.

I recall a senior member of the Justices' Clerks' Society once asking from the platform at a conference: 'What business is it of a justices' clerk to be involved in sentencing issues?' I believe the question was aimed towards the few of us who were dabbling in that field. Some of my published articles would have been a likely consideration for anyone making such a comment, and I had been involved in various national conferences about sentencing. I have heard similar remarks from time to time, but the narrow appoach has become more and more a minority viewpoint.

Recently, a president of the Justices' Clerks' Society spoke of the need for wider involvement in his annual report. What was being said in the past was that it is part of the job to advise on sentencing, but not to have a view on which way sentencing should go. This is quite correct so far as individual cases are concerned. But in more general terms I find the two aspects inseparable, not because I want to influence people, but because you cannot understand what sentencing is about unless you set it in a dynamic overall context. You cannot have a view that a particular disposal can be used in certain circumstances and then say to the magistrates: 'Well, cheerio. I will go now. You know the law.' Neither can you say: 'The advice of the Court of Appeal is this', because on most occasions it will have said something different on another occasion, possibly in relation to an offence of almost identical circumstances. Sentencing advice is not simply about being able to say: 'You cannot do this, because the Court of Appeal has said that you must be heavy on a particular category of offence.' You have to distil from the general background of sentencing law and practice those issues that relate to the decision which is about to be made. Things need to be opened up by drawing attention to the issues and to how different people have tried to resolve them, including, of course, the Court of Appeal. I have heard clerks say: 'The advice of Court of Appeal on this is so and so', and it then becomes a closed book. Often a striking statement or rule sticks in the mind and there is a lot of old case law which seems to have survived in this way. But a thing which comes over forcibly in recent years is the appeal court's own doubts about whether it was right in the past. Frequently, you will find a judgment in which it is urged that we move off in some new direction, or where the judges take what seems to be an unconventional course. This is not done without considerable thought, in a careful and considered way and often against a background of previous questionable approaches. Sentencing has not achieved what people may have hoped for over the years, and people are saying so.

## Parable of a Chief Constable

For several years, A3 was the chief constable of a county force from which he retired in the early 1980s. His reputation as an outspoken chief officer was still well known years later to many practitioners. He had few qualms about crossing agency boundaries, either within his force area or across the country, in his efforts to influence policing policy.

If you wish to seek the prevention of crime, you have to have constant communication with the public. Many of my critics thought I did this for self-aggrandizement, and that always being on television and radio or giving inteviews to the press was for my own personal prestige. It was nothing of the kind. It was the only means by which I could communicate what the force was trying to do for the public. This dialogue between myself, the force, and the public was critical to the ideas of community policing.

I had to sell these policies. Instead of just implementing them within my own jurisdiction and being satisfied with that, I began to write and publish and to talk publicly, particularly at universities. As you are not supposed to go from one police force area to another and hold meetings with people and talk about policing, I argued that you can go to universities because they are neutral territory. This drew me into conflict with my colleagues, the chief constables, who saw me as some kind of big-headed nuisance. I realized that there was animosity creeping into the business, which did not really surprise me, and it certainly did not deter me.

A3 began to feel increasingly out of place within the police service:

Without an understanding of human behaviour and the history of people, you are not going to do the job so well. At the end of the day, I had done so much reading that it rather made me an oddball in the police. I began to see myself as being unusual and, in a way, isolated.[9]

When you finally get to the top, all that experience and those influences should have produced somebody who is going to make a positive contribution. Anyone who is given a position of responsibility in an organization and who seeks to reform should strive to get a better relationship with the time, because each of us lives within a certain time scale and there are things that are right for that time. They may be wrong later on, and they may have been wrong if they had been introduced earlier, but they are right for their time. I have been accused of being ahead of my time and one can get too far ahead. It is rather like leading a battalion of troops through a wood, and you get so far ahead of them that they have lost you and you have lost them. You look around and they are not there. Now you can do that, and I suffered from that a bit I think, or they suffered from it, because they saw me disappearing over the hill. Where's he going? I was reforming and influencing their thinking, and you can be too far ahead. I suspect I did get too far ahead, but really that was just too bad and I do not have any regrets about that. They should have run a bit faster.

One senior police officer, A1, was an especially acute observer of A3's period as chief constable:

Looking at what one's personal objectives are and managing one's presentation, so that one is able to influence may mean compromising or concealing one's views until the timing is right. One has to take extreme care to maintain one's credibility if one is going to achieve the kind of change which hopefully will be sustained in the organization. There is the moral dilemma as to whether you pursue your views all the way, with the danger of becoming no longer credible and therefore achieve nothing, or compromise and go only so far, but at least forward. At the time, he [A3] was very much around and it was interesting to listen to what he was talking about and watch what he was developing—indeed, how it ultimately led to his almost alienation from the police service. One very quickly recognized that there were boundaries to acceptable challenge. In the light of his experience, and indeed of one or two others I had observed, I decided that one has to tread a fine line to retain credibility without compromising one's views or values too much. Ultimately you are being political. Perhaps it sounds scheming, but it is no use adopting a particular stance which leads to your being totally discredited as an individual and to what you have worked for being cast on one side. This would not be because there is anything fundamentally flawed with the idea as much as with the fact that they decided not to take you seriously.

Community policing was a main point at issue, which was allowed to go so far until it started to seriously challenge traditional police organization, in terms of the emphasis placed on enforcement and hard-line policing. Because [A3's] concern for it was sustained for such a long time, it started to soften his credibility within the organization. It moved him away from being the tough chief constable to being almost the director of social services. You cannot identify the point when it alters. It is a changing process, but it is not a matter of today he's OK and tomorrow he's not. Nobody says he is the best thing since sliced bread today and tomorrow says he is the worst thing. It does not happen like that, but somehow something happens within the police service where comments, rumour, and innuendo steadily become the collective. The individual sustains the outlook throughout whilst this process develops. Nobody stands up and says: 'We're going to discredit him today.' Somehow what is happening goes beyond the bounds of acceptability, and a few people will say: 'Well this is more than we are prepared to accept.' And it grows and grows and grows.

A3 gave evidence to the Scarman inquiry into the Brixton disorders:

Two or three years before the Brixton riots in 1981, I had written that unless something were done we would have riots. I did not think there was much doubt about that, as I told the Home Secretary and the Commissioner of the Metropolitan Police. But these were just the musings of a county policeman, not to be taken too seriously, but which in the end were vindicated. That was very important because it meant submitting evidence to Lord Scarman which was diametrically opposed to the Association to which I belonged. In the end I had to move outside my club and stand up and submit my own evidence. I thought it was important to state some views that had not been put, and that, unfortunately, alienated people from me. I drew attention to the hugely difficult job of the police. I was not critical of the police but, by implication, might have been critical of police leadership. Lord Scarman is supposed to have said that it was the most constructive evidence that he got from the police service.[10] But it was submitted at the price of disdain and unpopularity.

A1 on A3's fall from favour within the police:

His [A3's] separate evidence to Scarman was the last straw, but the processes, I suspect, had started long before that. It was just another nail in the coffin, but quite a significant one. There were landmarks that occurred on the way and it steadily got worse. At one stage, if you went to Bramshill and cited all his articles and books, that was good. Almost overnight, it seemed, after this process had taken place people avoided using him as a reference. Quite incredible. It indicated to me that if you are going to achieve anything then you have got to make sure you are not actually squeezed out like that, which is effectively what happened. It was terrible, not so much for him, as bad as that was, as for the service. They lost out there. But I cannot just blame the people in the service who decided to do that. He perhaps could have recognized the signs as well and managed it. I am making that judgement in hindsight. Perhaps he did not know it was going on at the time.

Did this discrediting of A3 give comfort to some parts of the police service? A1:

I doubt the process is as tangible as that. But there is probably an unconscious positive response, where nobody will actually say: 'Well that's good. He's out of the way now. We can concentrate on buying more plastic bullets.' It is much more subtle than that. How do you manage yourself through the system, and does it mean compromise or concealment? I see compromise or subtlety as being quite acceptable if the greater good is to achieve something medium- or long-term. Otherwise one could, if one was very blunt or lacked subtlety, get locked into a position of no credibility and never have the ability to influence.

Three further perspectives on A3 follow. Two senior police officers discuss aspects of the strain between him and the police service. Finally, a chief probation officer recalls his work with him.

A4, a senior police officer:

He was the commandant when I was at the College. I agree so much with his ideas, but he does manage to upset police officers. If he had swum a bit more with the tide instead of suggesting that he was swimming against it all the time. This was a time of change, and a lot of people were talking about the same sorts of things as he was but using different words. I agree with all of what he was doing, but not the way that he sold it as a piece of Holy Grail he had discovered and that he had the perceived vision. He was not terribly tactful about the way he talked about what he was doing, because a lot of other chief constables were doing similar things in their communities although perhaps not working quite so closely with their local authorities. They were talking about quality of service to the public and working with the community and listening to what the community says, and being more sophisticated about looking at where crime happens and how you can divert people from crime. He could have done much more as a pioneer within the police service, but almost wilfully suggested that he was out on a limb by himself.

A5, a senior police officer:

I like a lot of what he says although I find some of it, such as his grand design, quite irritating. He wants to structure and formalize matters, but a service such as the police actually needs to be more flexible. It is all very well saying it is a community council, but it is very evident there are people who do not participate in such forums who are vitally important to the end result. It is not the panacea. You have to think in terms of society being a network of groups, and not just of little groups of representatives.

C2, a chief probation officer:

He was refreshing, and the nearest thing that I ever found in the police to a philosopher-king. He talked, in visionary terms, in the most articulate way. He had a humanitarian approach, was a liberal in terms of his thinking and somebody who was not fettered by his job. He was always prepared to look at other opportunities, adopting a holistic approach towards crime. He did not see crime within the narrow setting of criminal justice, but saw it in its local roots. He appreciated that the answers lay beyond the police and the probation service in having communities which were reconciling and which offered opportunities to offenders. He also had the ability to hand-pick individuals in the force, sometimes fairly unlikely people whose careers perhaps had been rather dormant, but he recognized that they had the gift to follow and to inspire other people.

A3 joined the police service in 1946 and in due course was commandant of the Police College at Bramshill and an assistant commissioner of the Metropolitan Police before becoming a chief constable in the early 1970s. In accordance with the convention adopted for this book, he is not named, although his identity will be generally obvious. As he had no objection to being identified, reference may be made to how he was viewed by several chief constables interviewed by Robert Reiner in the mid-1980s.[11] A3 was one of the most outstanding criminal justice practitioners of his generation. He retired early from a police service that was unable to accord the same legitimacy to him as an individual as to the philosophy of policing which he championed. While it is difficult to extract general lessons from A3's career, the Credo Three practitioner has to worry constantly about retaining credibility within the organization without letting go of the core values that constitute his or her working philosophy.

## The Probation Service and Central Government

Political skills are a requisite not only within the agency but also in dealing with external bodies, including central government. The following extracts specifically refer to the changing relationship between chief probation officers and the Home Office during the 1980s. The intensity of the pressures from the centre was greater for the probation service than for other agencies and had two main threads. First, the Home Office assumed a much more centralist role, as exemplified by the Statement of National Objectives and Priorities (SNOP) and a series of guidelines on practice issues. Secondly, the government identified the probation service as being central to its concept of 'punishment in the community'.[12] The debate that followed extended beyond the choice of language to encompass fundamental values and beliefs.[13]

The passages that follow describe opportunities at the local level to exercise the discretion required to shape policy and practice. Initiatives of this kind arise, in part, from recognizing the plurality of objectives within central government and, where appropriate, making alliances with particular groups or individuals. For chief officers the skills include being able to distinguish political flag-waving from substantive proposals and also being alert to limiting the damage that public policy statements might have upon the morale of staff.

C1, a chief probation officer, recalls his promotion to deputy chief probation officer:

The job was a political one in many ways, and produced a lot of long-running and quite uncomfortable situations between ourselves and central government. This was partly about moving further and further from fieldwork practice and more and more to policy and central government initiatives. My life was much more dominated by battles with central government over a whole range of things than it ever had been previously.

I learnt that an individual area of the probation service had much more freedom than it realized despite central policy directives. There was some quite valuable learning for me about opportunities and areas of discretion. The Home Office's statement of national objectives and priorities was very firm that through-care should take a lesser priority.[14] We publicly stated that the Home Office had got it completely wrong, and that their lack of investment in decent through-care made after-care very much more difficult. In direct contravention of the national objectives and priorities, we continued to build up specialist through-care units. We took our committee with us and said: 'You must realize that if you take this policy decision, which we are recommending you to do, you may well have the Home Office and the Inspectorate down on you like a ton of bricks saying, "Why are you doing this? You are out of step with everybody else."'

After the argument we made it clear what we had done and we also did the same thing on day centres, when all around were producing more and more interventionist day centres. We were in an area that was rich in community alternatives and believed it was far better not to run specialist probation service initiatives, and far less interventionist to try and fit people into ordinary facilities in the community, even if it did not give the courts as much control. Our area had been under pressure from the Home Office for a long time to run its own day centre but always stood firm, saying: 'You've got that wrong too.' The point about locally organized probation services is that, while you are still accountable, you do have much more freedom because it is a lay committee. If individual probation committees, with their chief officers, take things by the scruff of the neck, constitutionally and practically they have the power to interpret criminal justice legislation very much how they think they should.

C4, another chief probation officer, comments on the government's plans for the probation service, and the implications for the underlying values of probation work:

I have worked in central government and seen how policy develops, and the mixture of political and professional imperatives, and can sail a bit with the wind and try and see those things that have a political flag attached to them, and those things that are about good practice. I try and put that over to staff. Knowing how keen some Home Office officials are on good, sound ideas properly worked out and resourced, I promoted our day centre work, crime prevention, and hostel programmes. I am happy to have a curfew in a hostel and to ensure that somebody goes to a day centre for eight hours or does their community service and attends to it in a gruelling way, and it has an impact. They like and will respond to that. But there is a streak around now, in terms of punishment, which is becoming a bit repetitive, and must not be bought for the sterile nonsense that it is. The latest research shows that punitive and restrictive measures alone do not work. What does seem to work is when there is, in a structured way and sometimes with constraints, a positive relationship, valuing people for the whole persons they are, and tackling the real, sometimes crying, personal and often social circumstances that they are caught up in. I told the Home Office that I do not want to know about punishment, which is about pain, negative restrictions, and silliness such as curfews and electronic monitoring.

C5, a chief probation officer, on attempting to work with, rather than against, the Home Office:

Some staff want to set up the barricades, but it is better to be in there and influencing the change. The best way is not to get into a battleground, but to take opportunities to influence. Over the years I have tried to have good links with the Home Office, and have taken opportunities to participate and influence the general direction, rather than just be negative about it.

I try to build up relationships with the different people at the Home Office, and have had quite a lot of contact with officials and certainly try to influence thinking. It is very important that we work together because what I do find difficult is the

Home Office view that they know best, and how it should be done. The Home Office are determined to take a very central approach, and therefore it is much better that we work with them at it, rather than just taking what comes.

## Supports and Supporting

The two-way process of giving and receiving support involves both individuals and groups within and outside criminal justice agencies.

E3, a senior Home Office official:

> I try to make myself accessible and give support to those who depend on the Office in one way or another, or to those who are trying to do things in difficult circumstances, where support or guidance, or often just encouragement, may be helpful. That goes both for people outside the Office and for colleagues. It is a matter of being alive to wider issues and principles, and of drawing other people's attention to them in ways which are supportive and help to generate a longer-term sense of direction and purpose. It is part of the wider duty, which we all have, that goes beyond satisfying the immediate practical or political needs of the moment. Ministers themselves will rapidly suffer if it is neglected.

Support is often found within the agency. E1 found that his fellow justices' clerks were more likely than not to welcome new ideas and innovation:

> People within the system were saying something at last, that things have livened up after all those years. When I first became interested in sentencing issues the reactions were not always favourable. There were some quite hostile seminars at the start of the juvenile justice initiative and critical letters occasionally, including strongly worded ones. A casual PS on a letter from another justices' clerk still sticks in my mind to this day: 'I suppose that by now your custody-free record has gone and you have locked up all those little buggers.' More recently, justices' clerks as a whole have supported the expression of ideas, even if not necessarily agreeing with them, and I think this has encouraged people to be more prepared to do this kind of thing

and become involved in wider matters and initiatives. With some people, confidence can stem from knowing that such involvement is recognized and acceptable. If the shape of what is happening changes, people start to respond, and start to change their ideas.

Networks of like-minded practitioners may be an especially powerful source of support, and many often do much to counter isolation. Such networks, as reported by the Practitioners, appear to be very informal. For example, A1 notes the existence of a network of like-minded officers, 'quality thinkers', within the police service:

> There are other officers of my rank in one or two forces who have a similar outlook but they are relatively few in number. We have not formed a club as such, but it is refreshing at times to meet these people. Keeping in regular touch with them is too strong a description, but there is an element of mutual understanding between certain police officers which comes about through contacts with other forces or at the Staff College. There are some officers who are happy to just carry on, but on every course or in every context you come across people with similar outlooks, who recognize problems within the organization and want to bring about change. Very quickly you exchange views. We are not a club, and it does not have a name, but there are like-minded people in the force up and down the country, and if I wanted to make contact with somebody in an area where we feel that there is a need for change, then I would be able to contact them. I probably know in every force at least one, if not more, who have this willingness to look afresh at things to see if they could develop the organization. We are talking about individuals with a desire to bring about change. In organizational terms you would call them 'change agents'. Invariably, they are recruited into development departments. They are used to develop, but they are kept within reasonably defined controls. It is to do with developing and with thinking. I stop short of saying 'thinking policemen'. All policemen think; it is the quality of their thinking that matters.

D5, a prison governor, identifies a rather similar network within the prison service:

There are three or four of us colleagues in the service who identify with each other in the way we examine the issues, and who are probably identified by others as people who ask awkward questions, both of ourselves and of the system, and who look for a theoretical framework within which to work. We have each coped in different ways. We meet informally for discussion and stretch each other intellectually. We are now in positions where we can influence the development of the system. We are broad contemporaries within two or three years in terms of service history and in terms of age. We are all in our mid- to late forties, children of the 1960s who probably also have had the confidence of prior experience before coming into the prison system and who have total commitment to it, but do not feel that it is a restrictive loyalty. We have done other things, and if pushed we probably could do other things again.

Has the evolution of this informal group played an important part perhaps in keeping such individuals within the service?

I have never thought of it in those terms, but my immediate response is that it probably has. We each have to find our own way of coping with the pressures of working in the prison service. If I did not have this intellectual stimulus, I do not think I would survive at a purely pragmatic level. The stimulus goes beyond Scotland, and I would identify a couple of colleagues in Northern Ireland and probably several in the English Prison Service who meet from time to time—not by design, but at conferences or courses—and who are all of a similar mind. I have developed an international awareness through visiting America in the mid-eighties and I have maintained contacts there. It is all of those strands which allow us to continue to work within what could be a very restrictive pool of the Prison Service.

On his initial prison service staff course, D2 recognized four or five colleagues who shared a similar set of values:

Out of the thirty people who trained with me on that eight-month staff course, there were four or five with a rather similar philosophy to me, also with the same way of presenting themselves. It was from people like that that one drew support. Only one of the others is still left in the service.

One governor, who is a personal friend, thinks what I write in journals is a waste of time, and says: 'You know, there are other people who can write this. You should get on with governing your prison.' On the other hand, there is a very senior official who would say: 'This is exactly what we want, a thinking governor who will publish.' He is a member of the Prisons Board and has always given me a great deal of encouragement in that direction.

I get a bit of support from the organization, but an awful lot of it comes from outside. In the academic world there have been people who have known the kind of work that I have been doing. One such person rings up for bits of practical advice about prisons, and has been supportive and has encouraged me to stick with it. The support tends to be from the academics and the practising lawyers with whom I have contact.

At a dispersal prison where I worked, we had some Irish Republican prisoners who were contesting all kind of civil liberties issues. When I arrived there, most were represented by one solicitor, and he [E4], the prison, and headquarters were at loggerheads over various matters. Some of his letters bordered on the abusive and the prison's replies were often unhelpful. I wrote to him: 'Look, we have rules too. I will try to make sure that your clients get everything that they are entitled to. The problems we have could best be sorted out over a couple of pints when you next visit one of your clients.' He replied that he would like to do that and we got to know each other really quite well. The spin-off for the prison was that he could now say to his client: 'You may not be getting what you want straight away, but they are working on it.' An example might be one of his clients who at that time had seen him about some matter, but during the interview must have said that he had given his watch to the chaplain to get it mended a long time ago and it had not been returned yet. He [E4] wrote to the governor: 'By the way, my client has asked when will his watch be returned to him.' A letter went out from the particular wing to say: 'Your client had no permission to discuss the matter of his watch with you. We will reply to him when we can tell him something.' The solicitor was asking a very simple question, but at that time Standing Orders meant that a prisoner had to express his grievance inside the prison and have it investigated first. I tried to implant a more reasonable and professional correspondence and to enhance it with a bit of personal contact. It helped and allowed his clients to see management, and the officers with whom they dealt on a daily basis, in a different light. A degree of trust became established.

E4, a solicitor, recalls his working relationship with D2:

He [D2] has done an awful lot of good work inside the prison system, at least making them aware that there is something outside. He is completely unusual and I have not met his like. They would not come near me because I already had a reputation by the time I started to go there. I was a 'trouble-maker', and they had instructions from head office that, 'if it involves him [E4], shunt it up here: do not touch it yourselves—he is trouble.' He [D2] was the first guy I came across who was prepared to stick out his hand and say, 'Hi, what is all this about? Talk to me; I want to know.' And he was the first man I came across who was able also to say, 'Well no, actually you have to look at it from our point of view. We have a management problem here. It doesn't mean to say that we've got it right, but we do have to manage and we can't simply surrender control of the prisons to the prisoners or to lawyers who sit outside.' He brought a perspective to it.

D2 speaking again of E4:

He has always offered a great deal of support. There was a time when I thought of getting out of the service. I applied for [another] job and did not get it, but at that time I would have left. He was very supportive at that stage and would say: 'What are you getting out of it for? The service needs people like you.'

Prisoners were supportive too. Something that I treasure, and have had framed, is a farewell card from all the prisoners on a wing, with some very positive comments on it. You should never underestimate the support prisoners can give, if they see you acting fairly. You come back to natural justice and being fair. At one large local prison, and primarily in the women's wing, support often came from the prisoners. Once on my night round I found a man hanging from his bars. Professionalism took over and I alerted the right people. The man was cut down and we thought we had saved him as his colour came back to his face. But we had not. Thereafter for me it became a management problem. I had to make sure the right people were informed, that they came in to the prison, and it all worked like clockwork. The practice of the officer who had been patrolling the landing had been a model of what it should have been. He had checked the man at the right time and in fact had checked him just before my visit to the wing. But he was in pieces because he believed it was his fault since he was in charge of the landing. I spent a lot of the night making sure he was OK and making sure other staff were OK. The prisoner was notorious, and the media rapidly found out about the suicide. In the early hours I found myself giving a radio interview over the telephone. Towards morning I was thinking: 'Well, who is looking after *me*?' I was upset as well, and I said to one member of staff that nobody seemed to bother about what I was feeling. She said: 'It's your job to manage the situation. You're not supposed to have feelings now. Give it a day or two and you'll get a staff response'—which is actually what happened. But the only person who immediately demonstrated some kind of care was an IRA lifer. At about 7 o'clock in the morning I did a full round of the prison and went into the women's wing. She had heard the news on the radio. She walked across the ground floor and put her hand on my arm and she said: 'I wouldn't have your job for all the tea in China.' That was very reassuring and very moving.

Looking for support from outside the agency and sometimes from beyond the criminal justice process may also be critical to the Practitioner's professional survival.

C2, a chief probation officer:

I knew my approach was different but it would be arrogant to presume it was better. If I could not find the mental support or stimulus I was looking for within the department, I looked outside and found like-minded and kindred spirits from the council of social services, mental health departments, or anywhere I could find it, and sometimes from the clients themselves. A lot of the time they provided an awful lot of inspiration and taught me a great deal about the job. My greatest teachers must be the clients and their families.

E3, a senior Home Office official, reached out well beyond the confines of his own department:

> I started to take the world outside the Office seriously from the time I went to the Prison Department in 1970. That was the period when I began to value particularly what voluntary organizations and academics had to say to us. From about 1980 onwards I tried consciously to build their contribution into the job. I found that those external contacts with the operational services, voluntary organizations, the academic world, and sometimes journalists were always stimulating and productive. I hoped to listen and sometimes to influence. Communication between the Office and the rest of the world is very important. In the sixties the Office suffered from its lack of openness and the sometimes not very well informed criticisms of the way it went about its business. Some of the same criticisms could still be made, but the Office was very much more open and communication was much freer in the eighties—and I hope it will become still more so in the nineties—than it was twenty-five or thirty years ago.

D4, a prison governor, confronts the natural isolationism of the prison:

> If the prison is not pointed outside, you will have an incestuous sort of existence. Prison is not an island or a planet thousands of miles from anywhere and should be part of the community. For every prisoner inside there are probably a hundred people outside, between his relatives, workmates, and the guys who drink with him. It is essential that people like me should keep their feet on the ground outside the prison wall. It is important at the personal level too. The more that people outside are taking an interest in what we are trying to do, so you get sustenance from outside.

## Inside–Outside

The Practitioners had made a choice to work inside rather than outside the formal apparatus of criminal justice. The restraints and opportunities arising from this choice mean that questions as to the relative effectiveness of operating inside or outside the agency are never far from the surface.

E5, a magistrate, on being influential from the inside:

> You have to make up your mind whether you can best pursue
> your attitudes within an organization or outside it. It took me
> about three months to make up my mind whether I would
> go on the bench or not. I decided then that perhaps I could
> be useful within the system, and the same applied with the
> Magistrates' Association. When I was elected chairman of
> the Association, I was concerned as to whether I would be
> hampered in my attitudes of pursuing interests that I thought
> were important, in that the duty of the chairman of any body is
> to fulfil the decisions of that body. I have sometimes been
> viewed with disfavour. It was a great surprise when I was
> elected chairman, and it was a pretty narrow thing. At one time
> I was determined not to accept the office of chairman of the
> Association, because my views have often been a bit more
> liberal than the average. People thought I was too ready, as
> chairman of an important association, to pursue a reasonably
> liberal course in representing the views of the Association.
> There are a lot of people on the less thinking wing of the
> Association who would say: 'Oh well, the chairman's saying
> that, we'll go along with it.' I found that I was in a better
> position to influence opinion by being within the organization
> than outside it.

E6, a recorder, emphasizes that, as far as she is concerned, being a
member of the Bar does not permit one to do only certain types of
defence work:

> One ought to stay in it. I very much disapprove of people who
> are in the profession and refuse to do certain types of work. I
> strongly disapprove (it is of course, against the Bar Rules, but it
> seems to be ignored) of people who say they will not defend
> rapists, will not appear for landlords, or will not prosecute,
> because they do not like the system. If you do not like the
> system, then get out of it. If you are in the system, then do it to
> the best of your ability. Otherwise you are leaving prosecution
> to people with an approach you might not approve of.
>   If you are reasonably good at doing what you do, and you
> believe that you do do a decent job, then stay in it and do it. If
> you have a chance to move things in the direction that you

think they should be, then stay in and do it. There is no virtue in washing your hands and saying: 'Oh no, this is not for me.'

C4, a chief probation officer, also seeks to avoid extreme postures:

It is how you try and stick in there and think maybe there is some influence you can have. I am a middle-of-the-roader; the system is not all bad, but equally it is not all good. I get angry and disappointed with the system, but I am in the business of trying to change it and to have an impact, and if I just go out on a limb and make a protest, I will have made the protest. But get stuck in and do something about it.

A good example is the work we have done with the judiciary who have had a pretty high sentencing rate. We have set out to get to know them, but also to tell them what we do. We have had High Court judges asking for our sex offender group specialists to tell them what they actually do, and then subsequently making orders. We have taken judges out to community service placements, and I have had requests to do similar things for our day centres and hostel work. And the *pièce de résistance* was being asked to talk to a sentencing conference recently on their sentencing practice compared with other areas. If I had said I had *wanted* to talk about that, I would have been reported to the Home Secretary.

For another chief probation officer, C3, the inside–outside line is sometimes difficult to define. Ultimately, her loyalties extend beyond the probation service:

In terms of speaking out on high-profile issues and articulating them, I have often found myself in a very ambivalent position. The probation service has failed to differentiate between what is a proper compaigning role, and the particular role of the voluntary sector to ginger us up, and to not let public-sector organizations like our own get complacent or feel that they have cornered the market in views about what is happening in the criminal justice system. It has often been quite a tense role that I have pursued, and people have called my personal loyalties into question, because I also have a role within NACRO and other voluntary bodies. My commitment remains to progressing work with offenders and achieving a more enlightened approach to criminal justice matters in the system in this country. I have no regrets about channelling my efforts through the probation service, but my loyalty and my commitment to that kind of objective goes beyond my preparedness to fight to maintain the status quo in the probation service.

This chapter has addressed the inherent tensions that arise between the Credo Three practitioner and criminal justice agencies. Occupying a senior position and enjoying a good reputation among colleagues are usually preconditions for being able to influence the course of events. Effectiveness in this respect means steering a course between co-option and isolation. For the Credo Three practitioner, such a way forward will often be highly taxing of resourcefulness and stamina. By definition, the Practitioners who were interviewed for this study had been able to continue as 'insiders'. Only one of them, A3, had retired from his post as a chief constable a year or two earlier than he had expected, in large part because of the particular pressures described in this chapter. A3 also viewed his departure from the police within what he regarded as an increasingly uncomfortable political context. He stated: 'I left the police because it was drifting towards becoming a repressive arm of an authoritarian government, and that was not my idea of policing in England.' For the other Practitioners, the issue of resignation arose very infrequently. Two came close to resigning at an early stage in their careers, but, in the event, felt able to remain. However, the casualty rate among Credo Three practitioners may be quite high, as D2 suggested with reference to his cohort group of assistant governors. It would be useful to know much more about such people and the particular dilemmas and pressures leading to their departures.[15] The focus of this study is working within the formal apparatus of criminal justice, and the Practitioners' careers demonstrate that the inevitable tensions and strains can be contained without unduly reducing effectiveness. As the next chapter shows, for the Credo Three practitioner survival within the agency is merely the precursor for actively undertaking reform.

# 6 Ways Forward

THIS penultimate chapter considers the implications of Credo Three for criminal justice reform and, in particular, for achieving structural changes that might best enhance humane values and resist countervailing forces. Initiatives and pressures for reform take many forms and derive from a wide variety of sources. The legislature, government departments, the media, advisory committees, and pressure groups may all, to varying degrees, be involved with respect to any specific issue.[1] Criminal justice practitioners are often the principal agents for change, being able to encourage, facilitate, or impede the reform efforts of others. Indeed, to a very considerable extent, practitioners comprise the linchpin that determines the success or failure of any reform endeavour.

The first part of this chapter considers the structural prerequisites for reform, with a particular emphasis on the implications for senior management. The essential interdependence of the criminal justice process is reviewed in the second part. Taking this wide perspective of criminal justice arrangements and being alert to the connections across agencies and phases in the process emerge as central themes of many contemporary reform efforts. Finally, consideration is given to the issues that are likely to be critical for the 1990s and the early years of the twenty-first century.

## Structural Prerequisites for Reform

A basic requirement for humanizing justice is for the agencies to formulate purposes that extend beyond expediency and survival.

This starting-point is articulated by C5, a chief probation officer:

> What has been lacking is a clear direction for the service which staff can own and contribute to. We need an overall framework which, working to specific objectives, builds in the values that we are trying to achieve. The ethos behind the service is very

much part of that, but staff must be left to say what they can contribute to the organization and must be given the opportunity to act creatively and come up with the ideas about how they are going to do that.

The second basic requirement is for participatory management that provides the structure for innovation and the development of good practice.

C2, another chief probation officer:

> We must be more delegatory in terms of passing responsibilities to people in middle management or lower down the echelon and to rely on others to deliver the systems that we set in motion. Any good organization must let plans, activities, and ideals, in the sense of innovation and good practice, be developed from the local level. It must avoid a situation where headquarters tell and the staff do. It is far better to have a policy which allows people to expand and develop their ideas from a local level. It is very hard to achieve that in an organization (sometimes because of the pressures of change), but if you give people their play they will come back with imaginative ideas.

Applied to the prison, it is clear that participatory structures must embrace both staff and prisoners. Sorting out the respective roles of the prisoner and prison officer may have to occur on an institution-by-institution basis, and, once established, a participatory regime will require constant renewal.

D5, a prison governor, provides an example of this type of structural reform, suggesting that, if it is to be successful, practice may have to precede theory:

> The first document we produced was broadly a presentation, in service-wide terms, of the principles which we were applying at my prison and which go back to the tradition of group work in the Borstal system, and to the notion that the key is the relationship between prison officers and prisoners. You have to allow those two groups to interact, and we set up structures enabling the prison officer to make decisions which affected the life of the prisoner, in terms of dealing with requests. They began reporting on the prisoner for local parole purposes with the

prisoner being consulted about the various reports about them. It was a fairly short jump from that to the language used in the document. In some ways we started with the practice before we reinforced it with the theory. We were able to do it because it was a new establishment and we established that tradition. The initial response from the prison officers was that they liked doing this, provided they had the support of social workers and other specialist staff. The prisoner was prepared to participate, once he recognized that the officer actually did have some power of decision-making. They will never love each other, but there was a degree of respect of the other's position. Once it was established, it was easier to develop the theoretical basis, and our challenge now is to apply that principle more widely.

Most Practitioners had a vantage point which enabled them to appreciate the potential contribution that good practice has for policy. C4, a chief probation officer, draws attention to the dynamic and interactive facets of this relationship:

I have always tried to keep in touch with good practice. After the Second World War, the Chief Education Officer for the West Riding of Yorkshire had a great reputation for going round his schools and seeing the best practice.[2] I have always admired that, and I like going to see staff, sitting in on a sex offenders' programme, accompanying an officer doing a report on a mentally disturbed person, or visiting a hostel or day centre. When I am going to meetings at the Home Office, I ask staff what I can contribute from a practice point of view. Staff, who may be depressed about lots of things, seem to rise to the connection that I am constantly trying to make between their practice and what I am about.

Becoming a senior manager within a criminal justice agency inevitably involves dealing with competing and conflicting pressures. The personal role adjustment on promotion is often made more difficult by having to relinquish especially satisfying tasks. Furthermore, the new member of the top management team has to make a special effort to preserve his or her personal working ideology in the face of corporate responsibilities. The Credo Three practitioner's determination to retain and extend a strategic and reflective stance may also be confounded by an organizational ethos that gives pride of place to short-term operations.

D5, a prison governor, emphasizes the self-discipline that is required in adopting a strategic stance:

I enjoyed being an assistant governor and dealing directly with prisoners and with individual officers. Part of the discipline which I now have to impose on myself as a senior governor is to stand back from that and not become involved in the daily struggles—for example, when I am walking about the prison, being seen, talking to people, listening to people. I have to have the discipline, when prisoners come with individual problems, to say: 'You have a principal officer or assistant governor. Go and discuss that with them. If it cannot be sorted out they will refer it to me.' That is a problem which a lot of governors face daily. If a major incident occurs, then one goes into operation mode and becomes more directive, but the discipline is to stand back and take the wider view, which probably was not done by governors twenty years ago. They were autocrats who responded on a day-do-day basis.

B6, a chief crown prosecutor, acknowledges some of the complexities that arise:

I have always enjoyed a challenge in my professional career, which is why I have set up three offices with the CPS. I do wonder where the job is going now, when we are four and a half years into the service, and we are coming up to a stage when we are going to be fully lawyer-staffed. The hard work is setting something up. Thereafter it will run fairly smoothly, with just the occasional crisis. To some extent, on a personal level the challenge is going for me now, with the job. I sit at the desk churning out bits of paper and circulating paper that Headquarters send to me, whereas I do quite like solving problems. I miss the challenge of that. Crisis management is very easy. It is long-term management that is very difficult. With crisis management you just walk in; you see immediate results and immediate improvements, and that is good. Long-term management is trying to change opinions, personalities, and everything, and that is difficult.

C5, a chief probation officer, on becoming a senior manager:

> The biggest jump is between senior probation officer and assist-
> ant chief probation officer, in terms of task and role, and it took
> me quite a while to take that on board. I missed the client
> contact and was very conscious at one time not to take com-
> plaints too thoroughly, as I tended to see that as the one
> opportunity I got to have some contact with clients. I found that
> quite hard, but it is probably the post I have most enjoyed,
> because an assistant chief probation officer has a clear brief in
> terms of developmental role and looking more strategically at
> the service, thinking about how you can take things forward
> and enable them to happen.

A1 comments upon the implication of reaching the rank of assistant chief constable, and draws attention to the organizational
pressures to be seen as a 'hands-on' manager. He refers also to the
effort required to retain personal values on assuming corporate
responsibilities:

> One has moved from an individual perspective as a chief super-
> intendent to somebody who has to have a more corporate
> outlook as far as the organization is concerned. The role is
> significantly different. I found great difficulty in making the
> adjustment to the rank of assistant chief constable, where one
> has to work very hard to maintain one's personal philosophy in
> the context of the departmental responsibilities which in turn
> accord with the other chief officers' ideas.
>
> Police officers tend to be pragmatic in outlook and therefore
> favour 'hands-on'. The further up the hierarchy, the less there
> should be 'hands-on', and the more there should be a strategic
> outlook. I like getting 'hands-off', but the culture of the organ-
> ization does not like that: they prefer to see chief officers being
> very practical in their outlook. If one is not very careful, one
> can be discredited in what one is trying to achieve by being seen
> no longer to be a practical policeman. It is very frustrating that
> there is this pressure to be concerned with day-to-day issues,
> important as they are. But, in my view, there is not enough
> standing back, reflecting and thinking: 'Where are we going
> from here?' and devising strategies and plans to get there. It
> should be very much reflecting on what the organization is

doing, and developing it, rather than making routine decisions about things which have already occurred. It also creates problems because one very often finds there can be too many things in the fire. Top management should be concerned with setting the framework within which the organization will operate, communicating, and listening, and should not be involved in the routine day-to-day decision-making about events, which are basically fulfilled by constables long before they get to chief officers.

A major inquiry in which I was involved indicated to me that there should be a much more philosophical and reflective approach to the function than perhaps is traditionally associated with police operations. I developed the role of 'the managing investigating officer', a distanced, reflective, examining, and reviewing approach, which allowed the officers to get on with the investigation for which they were well experienced. At the end of the day, what matters more than what the investigation reveals is the way it appears to have been done and the perceptions that people have of it, whether through the media or through contact. As a police organization we are good, efficient, and we do get on with the job; but often we do not disclose how well we have done, either at the time or immediately after. I saw myself as an interface between the 'outside world' and the investigation, with some overall hold on the direction of the investigation. At Bramshill, there is a course for chief officers on major investigations and linked serious crimes. They still teach the traditional 'hands-on' role for chief officers, and I am convinced that is totally wrong.

C2, a chief probation officer, insists that vigilance is required by senior management if innovative members of staff are to be protected:

There has always been a constant tension for me between innovation and maintenance. At times I was extremely frustrated and irritated by the pace of change. You face those kind of challenges in any organization. Managers have a responsibility to identify people who will take the organization forward. It is very exciting to locate those people who sit on the margins of the service. I have found in my own career that they need protection, support, and encouragement because there are divisive influences in the service which will seek to undermine people who put forward innovative practice.

## Independence and Interdependence

During the 1980s, it became more commonplace for practitioners to think in terms of the interdependence of criminal justice agencies. In this regard, Britain was behind the United States, where in the early 1970s a massive effort was made by the Law Enforcement Assistance Administration to encourage collaboration among criminal justice agencies, largely by means of financial incentives, at the state and local levels of government.[3] Although it is sometimes held that criminal justice is (or should be) a 'system', this word is generally rather loosely used to draw attention to interconnections across criminal justice agencies.

But regarding criminal justice as a 'system' may distort reality by obscuring the divergent and competing purposes between and within agencies, the informal working arrangements, and the unanticipated consequences that frequently ensue. Furthermore, smooth synchronization threatens the independence of agencies, a vital safeguard of due process. As Paul Rock has succinctly observed: 'Independent interdependence is the weak force that binds the criminal justice system together.'[4] For senior managers looking across agency boundaries, a perennial challenge is to work creatively with the inherent tensions, affording legitimacy to both independence and interdependence.

The establishment in 1986 of the Crown Prosecution Service (CPS) in England and Wales placed in high relief the issues of independence and interdependence. From the start, much was made of the agency's independence from the police, but, as some of the following extracts show, progress in that direction was dependent, at least in part, upon collaboration with the probation service. The independence of the CPS and the loss of control over the prosecution process by the police were developments that were resented by many police officers.[5] However, it quickly became evident that the arrival of the CPS also had far-reaching implications for other agencies, especially the courts and the probation service. The direction and pace of developments have placed the CPS in a pivotal position and suggest that by the turn of the century the CPS, like the office of the public prosecutor in the

Netherlands and Germany, may have become 'the spider of the criminal justice web'.[6]

B5, a branch crown prosecutor, compares his relationship with the police before and after the setting up of the Crown Prosecution Service:

> Before 1986 there were never real difficulties with senior police officers, but there were cases where one would make suggestions to them and there would be a certain degree of unpleasantness. Normally this was resolved in a way which was favourable, but it could be a far more difficult process of cajoling and other sorts of pressure, and a fairly grudging reluctance, possibly at the end of the day, when the police took your advice. One did sometimes temper advice a little bit so as not to rock the boat. We did not fully appreciate that the setting up of the Crown Prosecution Service put us in an equal position to the police. They obviously objected to a considerable extent when they lost that power. It is astonishing how much the job has changed since the service has been set up. It makes you operate far better and more independently. On a personal basis, the additional status and responsibility combined with the changes has made me a more powerful arguer with the police. I am a different operator now to what I was five years ago, when I was little more than counsel with instructions going into court.

B2, a chief crown prosecutor, on the integrity of the confession:

> Since 1986 prosecutors have become more and more aware of the importance of protecting the integrity of the confession, and this has been encouraged by some of the more appalling cases that have hit the media. Prosecutors are particularly cautious when faced with uncorroborated confession evidence. Experience shows that courts are increasingly reluctant to accept uncorroborated confession evidence, and there will be occasions when a prosecutor will decide that this evidence alone will not justify a finding that there is a realistic prospect of conviction. It is not that we distrust the police, but there have been too many cases where the police evidence has had grave doubt cast upon it not to be particularly careful about accepting the uncorroborated evidence of the police regarding a confession. It is sad that one should have to say this, but we are more careful

than ever now. Of course, the introduction of tape-recorded and more recently of video-recorded interviews has helped enormously to dispel any doubts the defence may try to cast upon the reliability of confession evidence. The presence of defence representatives at interviews also helps to give added credence to the police version.

A1, a senior police officer, distances himself from criticisms of the CPS voiced by many of his colleagues:

> It is very easy to fall into the line adopted by many of the critics of the new Crown Prosecution Service, based on very practical experiences or disappointments in some instances with an organization which has had to grow from nothing. I prefer the prosecution process to be out of police hands. I am a supporter of the CPS. It is good for justice that there is another independent body. Generally speaking, I subscribe to the principle that the decision to prosecute is for a separate and independent body to the organization which has carried out the investigation. As the CPS gains more experience, so the service is getting much better. We have all had our difficulties, not least of all the police organization, in adjusting to this new body, but we are getting there together.

## THE PUBLIC INTEREST

A significant consequence of establishing the Crown Prosecution Service has been the renewal of attention to the concept of the 'public interest'. Public interest considerations, long acknowledged with reference to prosecution decisions, were given particular prominence with the publication of the CPS's Code of Practice.[7] Furthermore, although the public interest is most commonly articulated with reference to the Crown Prosecution Service, it is also relevant at other stages of the criminal justice process. The key sequential elements in operationalizing the public interest concept are: setting out the underlying principles, specifying the criteria arising from these principles that structure decision-making, and, finally, providing guidance and support to the practitioner in the exercise of these decisions. Each stage concept provides opportunities for the Credo Three practitioner to counter the preferences sought by adherents of competing ideological credos. Some of these issues are explored by three chief crown prosecutors.

B1, a chief crown prosecutor:

We should be looking for ways of diverting people away from the criminal justice process, which is a new exciting area for us all, and has yet to develop because it has just started. I look forward with optimism to the next twenty years or so to see just how far we can go.

Some of the ideas that I have now are more mature than they were before the Prosecution of Offences Act, but as far as the philosophy of prosecuting is concerned I have not changed. There is now a greater awareness of what other agencies are doing, and that we are just one link in the chain. The coming into existence of the Crown Prosecution Service has concentrated everybody's mind on the general area of public interest discontinuance which before 1986 nobody had really got to grips with. Prosecutors, of course, were aware of the Attorney-General's Guidelines of 1983, which directed chief constables to the idea of assessing what was in the public interest. There were occasions when I would look at a case and say, 'Well the evidence is there, but is it worth it? What are we going to achieve?' But there was no rubric to go by, although I felt uneasy as to what we were achieving. I never really thought to myself, 'Am I exercising the public interest criteria here?' People now have their attention directed to it immediately, rather than it being somewhere in the background.

B6, a chief crown prosecutor:

I am concerned, at times, that the CPS lawyers I meet from other areas are not switched on to public interest. I am very keen to get away from this view that, because a man is going to plead guilty, you prosecute him. I want every case looked at, and looked at with some detail there. I have some sympathy for people who are committing one offence. I think at times that the consequences of the prosecution are so out of all proportion to the offence that it is entirely wrong to give that man a criminal record, and it does not matter to me whether it is a juvenile or a young adult, or an even older person. Anybody can have a moment of madness, criminal madness even, and I do not think it's right to let that particular moment blight their whole career and their life. Public interest does not insist on prosecution. As a solicitor, I defended two or three women who were accused of shoplifting, first-time offenders, and they went through agonies waiting for the court case; they had gone on to tranquillizers, and it affected their lives, their marriages, their relations— everything. Irrespective of whether they did it or not, I am not sure the prosecution was worth that effect on human life, and I have always tried to steer away from what I thought were repressive prosecutions.

B3, a chief crown prosecutor:

Although no two people will articulate public interest criteria in the same way, it is important that we take into account the same sort of things. The Code makes it easy: you take into account the age of the defendant, the staleness of the offence, the likely penalty, and so forth. We are given the broad principles, the paint and the brushes, and it is up to us to use them to paint the picture that means something to us as people in the particular circumstances of the case. The day I am told that the public interest requires no exercise of discretion on my part is the day on which my position will cease to hold its attraction for me.

I am seeking to establish a professional environment where, if a member of staff has a problem, there is always someone they can turn to, talk to, and mull it over with. There should be no one who is isolated with a problem, for example one concerning the exercise of discretion regarding the public interest criteria. Problems can arise when a young lawyer who has just qualified is presented with the Code for Crown Prosecutors and told to apply it. We do ensure that our lawyers receive proper training, but you cannot rigidly train someone to apply the public interest. It is experience that is the main teacher, and experience that has to be shared. Mature entrants into the Crown Prosecution Service, who generally have more experience of life, probably find it easier to apply the public interest criteria than those who are straight out of university or law college. It is a matter of training and discussion.

An intriguing and unexpected aspect of the setting up of the Crown Prosecution Service was the opportunity for partnerships between the CPS and the probation service. In part, this opportunity arose from a new appreciation of the notion of interdependence of criminal justice agencies providing the basis of reform initiatives. Pioneering work in England and Wales by the Vera Foundation of New York was especially instructive in demonstrating that the provision of information by probation officers to the Crown Prosecution Service could reduce the number of persons remanded into custody. Under the Prosecution of Offences Act 1985, the responsibility for deciding whether or not to oppose bail passed from the police to the prosecutor. That bail information schemes, as they came to be known, could play a de-escalating role of this kind was recognized by the Association of Chief Officers of Probation and by Christopher Stone, director of the Vera Foundation's London office.[8] This approach to inter-agency collaboration was later adapted to the earlier stage of decision-making by the Crown Prosecution Service as to whether there were public interest grounds to discontinue proceedings.[9] The new partnership between probation officers and prosecutors has worked in the direction of de-escalation. However, partnerships across criminal justice agencies may have the contrary effect of weakening procedural protections and encouraging more punitive outcomes.[10]

B1, a chief crown prosecutor, on his surprise at the new relationship between his agency and the probation service:

> Prior to the Crown Prosecution Service being set up, I would have been surprised at the amount of liaison that there now is between the probation service and the CPS. Probation has always been, and with a number of people still is, associated with the opposite of what prosecutors are doing: with trying to make it better for defendants and being somewhat naïve in some of the recommendations put forward. In certain areas there is still a degree of antipathy towards the probation service. But there has been a growing realization that, if we are going to exercise any independence in the criminal justice system, we have got to have information. And when you look around at the sources of information, there has been just the police, which of

course is bound to be one-sided. The probation service fitted the bill because it appears to be best able to provide the information we are looking for.

C4 looks at this relationship from the vantage point of a chief probation officer:

> It has been an incredible learning curve, working with the CPS on bail information schemes. We are not there as social workers or the defence, trying to help offenders in that sense, but are verifying critically important information, to allow people to make decisions. We now have the opportunity to work with the CPS on discontinuance. With what we know about an individual's circumstances, we can advise the CPS to discontinue a case, whether on mental health or poverty or any other grounds. There are lots of controversies around, but these are a very exciting set of developments. Here is the probation service at the forefront of developments in the criminal justice system. We have to learn from the West Europeans, without doubt. Bail information has been a very good bedrock, and the research that Vera did was very helpful. Discontinuance work with the CPS will cement that, if the will is there and if the police do not get too scared about where we are in all of this. Probation officers will have to learn that the focus of attention is not always on the offender as an offender *per se*, and that we are working with people caught up in the criminal justice system at different points.

B3, a chief crown prosecutor, develops the point made above by B1 that one path to independence from the police lies in the new-found collaboration with the probation service:

> The birth of the Crown Prosecution Service has led to the realization by other agencies within the criminal justice system that they cannot act in isolation, and to a greater recognition of the need to seek partnerships in changing attitudes and the criminal processes. For example, in bail information schemes the probation service assists us in providing additional information concerning the background of an individual defendant that is independent of the police, so that when a CPS lawyer is making a decision as to whether or not to object to bail, he is better informed because of their help. We have created a greater

awareness of the considerable benefits which can be achieved as a result of partnerships within the justice system, and there are lots of people now wanting to have discussions and dialogue with us.

B3 also speculates on the possibility of greater collaboration, in certain circumstances, between the police and the CPS during the investigation stage. But moves in this direction may undermine the independent status of the CPS. This issue arises, for example, with respect to the setting up of specialist units to address offences such as drug trafficking and fraud.[11]

As for the future, one way forward would be for the CPS to have greater involvement in the investigation of crime. This is sensitive territory because it impinges on the responsibilities of the police. The police have to make difficult decisions, subject to challenge by the courts, for example on excluding a solicitor from an interview in the police station, or whether a particular prisoner should be further detained when the clock under PACE is running.[12] In my view, it would help the police if the CPS were involved in those decisions at the time they are made rather than merely dealing with difficulties arising in the prosecution of cases where sometimes we feel a wrong decision has been made. In areas where the police are vulnerable in their investigations, a CPS lawyer could assist in the decision-making process.

I would not want to take the decision to prosecute away from the police. I like this partnership, and I feel it is in the interests of society that the police make the initial decisions. Policy decisions as to the use of police resources and the general enforcement of law and order are properly within the control of a chief officer of police; but I would like to see the CPS involved in some of the very difficult decisions that police officers (who generally do not have our familiarity with case law and the way the judges are thinking) have to make in individual cases.

Currently the responsibility for making those difficult decisions is with the police. If the police approached me, I would have to say to them that Parliament has given them the decision and they have got to make it, however difficult. The relationship I had with the police before the inception of the Crown Prosecution Service meant that, if they had someone in the cells and had to make difficult decisions, they would slip into my office and discuss it with me. The decision would have been that of the police officer, but it would have been made after a full discussion with a prosecuting solicitor. Because we are now an independent service, with a statutory role, the police cannot so easily avail themselves of that advice, and we have to be careful not to tread on their toes.

That there was resistance within the courts to the CPS was hardly surprising. Informal traditions and unwritten rules govern the patterns of work of many courts.[13] One example is the court's discretion with respect to mode of trial, between the magistrates' courts and the crown courts, of 'either-way' indictable offences. To deal invariably with domestic burglary cases in the crown court, for example, would have an escalating effect on the use of custodial remands and on the length of prison sentences.

B1, a chief crown prosecutor, decided to confront directly a magistrates' court which had adopted this approach:[14]

> I place a great deal of importance on the mode of trial decision. I firmly believe, with regard to many offences that are being committed for trial, that it is not in either the prosecution's or the defendant's interest. They end up at crown court when they could be just as easily dispensed of in the magistrates' court. This brings us into conflict with certain magistrates' courts, which have their own particular policies. I have one magistrates' court which has a very firm policy that all burglaries of dwelling houses, regardless of the facts, go to the crown court. I am in some difficulty because there is some judicial authority for saying that should be so. But I do not believe it is so, and I tell my prosecutors that we ought to weed these out and make representations that they be dealt with summarily. There is always the difficulty that prosecutors might actually agree that the policy is correct, or that, over a course of time, they realize that they are banging their head against a brick wall and that all they are succeeding in doing, by making representations that the bench do not like, is becoming unpopular, so they resign themselves to adopting the policy. By means of certain initiatives, we have tried to get people away from that.
>
> I showed the bench all the figures and pointed out that they

were committing more people for trial than any of the other petty sessional divisions in the area, and said: 'There may be reasons for this, but it is incumbent upon you, the clerks, and me to think about it and to come to some conclusion as to why that should be so.' They expressed great surprise, because they had no idea that they were committing more people for trial. The more I can exercise some influence by going round having regular meetings with magistrates and magistrates' clerks, the more can come out of it.

## THE COURT AND THE PROBATION SERVICE

An especially striking development during the eighties was the growth of mutual respect and co-operation between probation officers and sentencers in both the magistrates' courts and the crown courts.

E5, a magistrate, places this new relationship within the new perspective of interdependence:

All these experiences have been cumulative towards my view of the totality and interdependence of the criminal justice system. If you recognize that, you are going to adopt a more liberal approach than you would if you view the bench in isolation. The concept of the criminal justice system has only recently become popular, of the interdependence of us all and how what we do in court influences others—in particular, the relationship with the probation service and the need for a constructive tension between us while clearly understanding the differences of our roles. The probation service suggests the means that are most likely to minimize offending, but the court has the wider remit of absorbing the anger and anxiety of society. In the magistrates' court we do little else, and sometimes help to minimize reoffending, but very importantly to ritualize the dealing with offenders to prevent individual reprisals. If courts are seen to be totally out of step with society, there is always this danger.

By the mid-1980s, the probation service had achieved some success in raising its credibility with the crown courts. The proportionate use of custodial sentences began to decline, and there were tentative indications of a turn in judicial attitudes, at least on the part of the part-time judiciary, namely recorders and assistant recorders.[15]

E6, a recorder, reflects on new attitudes within the Crown courts:

There are newer and younger judges who are more prepared to look realistically at sentencing policy, to understand that prison is ineffective, inhumane, and expensive. They are beginning to see that. There has been an improved relationship between the probation service and the courts; and in the magistrates' courts, and with respect to juveniles, it has been quite dramatic. As far as the crown courts are concerned, it is beginning to work, and in some places probation officers get on very well with judges and meet them regularly. There was a period when circuit judges developed a distaste for social work (in which they included probation officers), because some probation officers make unrealistic recommendations. There has been a real attempt by the probation service to provide better reports, and that way forward is working. If they get your confidence that the sentence is going to be effective, you have to be pretty stupid if you do not try it out. It is changing with assistant recorders and recorders, particularly those who practise in crime, who go in and out of prisons. They are meeting people all the time who have been in prison, who tell them they have been banged up for twenty-three hours, and they know what the conditions are, and do not close their eyes to it.

At a seminar for judges there was a presentation in the afternoon by the probation service. The reaction was 'What are they wasting our time with this for?' from a large number of the permanent judiciary. But from the junior end of the judiciary there was far more interest, and an acceptance that you really did need to know about this sort of thing. There has been so much talk recently with the Green and White Papers, and suggestions about changing policies and sentencing, that people are beginning to think they do have to look beyond the guideline cases.

## LOOKING UP AND TAKING A WIDER VIEW

This section concludes by returning to the emerging consciousness among practitioners of the interdependence of criminal justice agencies. The notion was very much to the fore in the late eighties, and multi-agency collaboration was regarded as being a cornerstone to efforts to divert persons from the courts, to encourage the use of bail, and in the development of community-based alternatives to custody. In the period since most of the interviews, the notion of interdependence has received further prominence. In 1989 the Home Office commenced a series of 'special conferences' for senior practitioners to encourage this wider view of criminal justice. Further support was given in 1991 by Lord Justice Woolf's report on the disturbances at Strangeways and other prisons.[16] The government accepted Woolf's proposal that national and local forums be set up, so as 'to promote better understanding, co-operation and co-ordination' by criminal justice agencies.[17]

B5, a branch crown prosecutor, expresses the new appreciation that effective reform may depend upon partnerships that extend across agency boundaries:

> It is exciting to feel that you have a role to play in the criminal justice system, whereas before you just did your job, kept your head down, and did not get involved in what was happening in the wide world. Now you can see ways in which the criminal justice system can be improved. Despite the pressures, the CPS has made itself open-ended, and this has liberated me to a considerable extent. I had become a little bit tunnel-visioned with the police telling me what to prosecute, but this new awareness of what the prosecutor can really do, and having seen what we can do nationally, has been very invigorating. I would like to see the movement that has now been established capitalized upon and expanded enormously.

This perspective is further developed by B2, a chief crown prosecutor:

> Almost overnight, we could for the first time impact on and influence the criminal justice system. That was certainly new. Some of my colleagues are not altogether happy with this prospect, but I welcome the opportunity to influence the system and I want to ensure that the CPS influences it in the right way. It is very important that we are informed and that we take part in the debate, because we have a tremendous amount to contribute. It is very tempting for many of us, who are lawyers and

not expert managers or academics, to retreat behind our case-work and say: 'Parliament has given us the responsibility for prosecuting and we should stick to that job. What we are is basically lawyers, and we should get on with the job we've been given as lawyers.'

Interdependence is also recognized as a central theme by some senior members of the probation service, such as C1:

This is probably the area in which I have changed the most over the last three years. I meet regularly with my chief constable, the chief crown prosecutor, and the justices' clerks. They are the key figures to work with. I am doing co-operative work with the chief constable on adult cautioning, and with the chief crown prosecutor on ways of identifying bail information. We are much more interdependent in the criminal justice process than I once thought we were, and also much more powerful if we have got joint schemes that work. The chief officers in the agencies are generally ahead of sentencers in terms of wanting some of the things that I believe in. The juvenile offender initiatives would have been impossible if we had relied solely on the co-operation of sentencers. It was only when faced with a group of committed chief officers that they decided to give it a try, and it is now well ensconced, and not likely to be shifted.

C3, a chief probation officer, notes that the particular onus rests with chief officers:

The outward- and forward-looking responsibility falls uniquely to the chief to carry. I feel quite positive about that bit, even on the days when I am depressed about White Papers and Green Papers and the other documents, because a momentum has been carefully worked away at which is about much more recognition of the interaction of parts of the criminal justice system on each other. It needs to be looked at in a much more coherent way, and it needs to be much less insular, too, recognizing the outside world.

C5, a chief probation officer, is concerned to discourage any isolationist tendencies within the probation service:

I do not see the probation service within a very narrow view of just supervising offenders, but look much wider than that in order to see where we can intervene. That has to be the strength of the service. We are one of the very few services that have links with so many different parts of the system. You can actually use that to the benefit of the whole. I welcome powers to contract out and grant aid,[18] because it is particularly relevant in areas like ours where we could actually grant aid to some of the small Asian organizations. The probation service went through a period of ten years or so when we developed our own professionalism and our skills at tackling offending very well; but we became very introspective, within a sort of separate organization. We have got to push out much more to working with the local communities and actually using local resources as part of the work that we do and must not pretend that we have all the answers.

I would like to see the probation service operating within the criminal justice system, both locally and nationally. There has to be a lot more commitment at government level to enable that to happen. I was appalled, at a time when we are talking about crossing boundaries and trying to work together, that the Prison Department set up a completely new structure which has no relation to any of the other agencies. If we are going to achieve these goals we must take a flexible approach, offering a service at each stage of the criminal justice system, and not just looking at working with offenders. Crime prevention is going to become very important to the service. It is very important that it is local, and, while some amalgamations may be appropriate, I do not see huge services being the answer to tackling crime in the local communities.

D5, a prison governor, decries the traditional isolationism of the prison system:

The dictum of the Wynn Parry report of 1959, that the prison service is *sui generis*, is still firmly held and is very important to a lot of people in the system, particularly in English prisons (where prison governors adopted that phrase as their motto for some time).[19] People jumped from that to saying [that] the prison service is completely unique and that any comparative study is irrelevant. But it seemed to me that a study of bureaucratic organization in general, and with reference to the prison service in particular, would actually be quite useful. One of the root difficulties which faces the prison service is that it is located in the mainstream of central government administration, rather than firmly in the criminal justice process, which is historically where it should be. The links between the prison system as part of the criminal justice continuum had been significantly weakened by the changed organizational structure, following the disbandment of the Scottish Prison Commission, and that contributed significantly to the organizational problems throughout the service.

## Challenges Ahead

This final section considers a number of themes likely to be to the fore in the 1990s. Firstly, there is the insidious issue of racism.

C4, a chief probation officer, on racist attitudes within and beyond the probation service:

It is looking both ways to the black community, and seeing how our work is perceived by them, and what you can do about that perception, and looking to the courts and prisons, the police and the CPS, and the other parts of the criminal justice system, and saying: 'What is happening here? These figures, those individual experiences that we see in court—what do we make about all that, how do we stand up and be counted on that?' We have tried to set our stall out and continue to work at that, and it is uncomfortable, both ways. The black community are suspicious, but, having worked with them in certain areas over a number of years, those barriers are breaking down. We have schemes whereby staff are appointed to work alongside probation officers doing reports, to work in court, to get the black dimension in perspective, to work with black women in prison, and we are also trying to do work with the judiciary and the magistrates, both on our committee and in training sessions and meetings. I was asked at a meeting of magistrates whether there was racial prejudice in magistrates and in their sentencing, and I said a resounding yes, and went on to explain why. I was very pleased when one or two black magistrates came up to me afterwards and thanked me, because there are still a lot of clerks and magistrates who disagree. Even more embarrassing, as we develop and promote our training and personnel equal-opportunities policies and get more black staff at various levels in the organization, I have been appalled at the way some staff have treated their black colleagues, and that must be rooted out. I am very concerned that it is very endemic.

A2, a chief constable, acknowledges the work that remains to be done:

> We are much more enlightened now than we were, but there is still a lot of work to do. There still are not many forces, and I include my own in this, where we have yet got a firm policy about racism. There is much statement-making and some real aspects of action, but there are not policy declarations as visionary points of focus. That is a piece of work that I now have to do. I tried to interest the Metropolitan Police in it from 1983 onward, and, although it was debated earnestly enough, it did not ever take root. It may do now with equal-opportunities policies. The backdrop of understanding is now much more receptive for statements of that kind.

E2, a justices' clerk, challenges conventional attitudes about race, including some of his own:

> All the institutions in this country, legal or otherwise, contain a high element of racial prejudice. I see it in my own organization, the people I work with, the magistrates, members of my own family, and myself sometimes. The hearts and minds for a multiracial society are only going to be achieved over a long period of time, of generations of work with schools and everything else. But we have got to remove some of the demonstrations of prejudice from institutions, such as the criminal justice system (even if magistrates do not recognize that discrimination exists, because they say the judicial oath prevents discrimination). The potential for discrimination exists in the criminal justice system because prejudice exists across the whole range of institutions and society in general. We have to monitor our results and build double checks into bail and sentencing decisions. Are we refusing bail because he is black, or because we do not know his background? We have to acknowledge that prejudice and the potential for prejudice exist. One of my battles will be to raise the issue of racial awareness with the bench as a whole, and to build it in as part of their checks and balances in their judicial decision-making, particularly bail, adjournment for social inquiry reports, reactions to recommendation in the

courts, and finally the custody and non-custody decision. I see young black men as potentially threatening, and I am sure that one would sentence accordingly, unless you pulled back and asked yourself [why].

E6, a recorder, on breaking down racial barriers at the Bar:

Increasingly there are black faces in chambers, and now the message has got through that you cannot turn people down because of racial prejudice, and that is slowly breaking down at the Bar. As for the criminal justice system as a whole, we have a long way to go. We start at the stage of arrest, and we know that young blacks are more likely to be stopped. They were more likely under the 'sus' law to be stopped,[20] and they are still more likely to be stopped and searched for drugs. It reflects prejudice that actually exists in society. There is no good closing our eyes to it; there is a lot of it about, and we will not cure that overnight. But we can make sure that, once they get into the system, we lean over backwards to be fair.

Credo Three values may become more precarious than ever with the introduction of cash limits, performance targets, and other modes of centralized control. For example, with reference to the police, crime prevention and community relations work is especially vulnerable to fiscal controls, which tend to be tied to conventional measures of efficiency.

A5, a senior police officer, argues that police officers adhering to these approaches are likely to become even more hard-pressed to defend them:

The challenge of the nineties is value for money versus quality of service. Providing clearly measurable results so far as policing is concerned compromises the notion of community policing. Quite often, those with elected authority are isolated from the needs of a small community, and you must also deal directly with the people concerned. To do that you are going to spend a lot of time, energy, and resources on communication, the results of which inevitably are immeasurable—for example burglary initiatives, with officers going round and persuading people to protect their property by crime prevention methods; or, in cases where people through lack of ability or resources are unable to do it themselves, actually arranging for it to be fitted for them and encouraging other agencies to participate. These activities are not clearly measurable. You can measure an operation on personal robberies on the top decks of buses. Someone might say: 'There is £20,000 for you to run an operation to catch the people who are doing that on the buses.' We probably know through intelligence who the youngsters are, and so we set up surveillance operations on the bus and perhaps on the individuals concerned to see what they are doing. The outcome will be the number of thieves that are caught. But far more lasting than that sort of operation, with the easy objective of getting those youngsters caught and sentenced (which has seldom if ever proved to be effective in terms of recurrence), is to work with youth clubs and schools to prevent them from doing it in the first place. This is resource-intensive, but it is far more important in terms of the quality of life for individuals who then work in this area than the other operation. But it cannot be measured.

A2, a chief constable, argues for a less 'self-centred' police force:

There must be a much greater awareness by the police service of the communities which it polices. It has got to be less self-centred than it is and we must try to put the centre of our activity out into the community. In order to do that, we should be prepared to look at our own methods of working with some detachment; we should try to understand that we do not have the monopoly of views of what is 'value for money' in policing. The community is paying for policing, and will have its own view of what is value for money. We must also make sure that we are less self-serving as an organization than we are inclined to be. I sometimes use the symbolic vignette of a police constable on duty and an elderly person waiting to cross the busy road; in the opposite direction, the chief constable pulls up in a staff car. The tendency would be in 99 per cent of the cases for that constable to turn to the chief constable to ask if he is being attended to. This is very flattering for the chief constable, but actually neglects the member of the public, whom both are there to serve. It is an image that I want to change. The police service spends so much time servicing itself and its own organization, bureaucratically and hierarchically, that we reduce the level of service we are able to give to members of the public.

Another major direction in which we should travel is towards greater openness and honesty. There are areas of our work which have to be secret. Terrorists and bank-robbers alike can choose the time, terrain, and method for their attack. The police should not feel obliged to reveal their methods of prevention and detection to a point where a further advantage accrues to enemies of society. Similarly, *sub judice* rules are prescribed for the proper avoidance of trial by media, and police officers should be duty bound by them. However, as a service we have been inclined to cloak ourselves in unnecessary secrecy, not least to avoid red faces when we have made a blunder. Mistakes are inevitable in busy, uncertain lives, and we should be more prepared to stand up quickly and acknowledge when we have got it wrong. I have certainly found that communities are much more prepared to give us the benefit of the doubt if we are prepared to be open and to help them to understand the pressures we face. It also makes them, in my experience, quick to sing our praises for all the good that our officers achieve.

To do that, there has to be a level of intellectual honesty and integrity. There comes a time when posturing has to stop. We have to sit down and work out a new compact with our communities. We must continue to hold the dirty end of society's stick—that is, our paid duty on behalf of the citizenry. But those same citizens can be trusted—must be trusted—to give a view on *their* priorities. I believe that, under the right circumstances, they can be persuaded to help us find, with them, common objectives and some sensible, practical measures to show whether or not we have been successful on their behalf. The Police Authority—as elected members and magistrates—have a vital, representative role in this co-operative effort. Although individual issues will sometimes spark off combative flashes, by and large we should not dissipate our energies in fighting with each other when there are battles to be fought in improving social justices, bettering education, and reducing the opportunity for crime.

B2, a chief crown prosecutor, suggests that the CPS seems likely to make deep encroachments into the responsibilities of not only the police but also the courts:

As prosecutors, we ought to keep our minds open to the possibility for change and allow ourselves to be influenced by others within the criminal justice system, who are contributing their own views and experience to this process of change, which is certainly gaining momentum. One would have been very surprised a few years ago to have been told that the prison population would dip as quickly as it has, and that two-thirds of juveniles and a fifth of young adults would be cautioned. We are just learning to acquire the ability to express our views as a national service. It is going to take a lot longer before we gain as much confidence as I would like, but we are beginning to see that we have a voice that ought to be heard.

But is it right for the police to caution two-thirds of juvenile offenders without us ever being invited to comment, or for an increasing number of young offenders between 17 and 21 to be cautioned without us being consulted by the police? In any event, the cautioning guidelines prepared for the police were issued before the CPS came into existence.[21] I am not persuaded that the police are necessarily always the best people to interpret the public interest, but that is what they are doing at the moment, and they are doing it in a number of informal ways with the probation service, through contacts with other agencies, and in a way that could not have been imagined even ten years ago. Some might say that the police are effectively usurping the statutory responsibility of the CPS in their decision-making. I am not saying it is feasible at the moment for the police to adopt a wider role, for all sorts of resource reasons; but I would be much happier with a system where all offences were reported to the prosecutor and the inquiry was conducted in his name, and the prosecutor decided whether someone should be charged or cautioned, or referred to another agency for help. The CPS has to decide whether it wants to accept a very narrow definition of its role, or a much wider responsibility in the public interest.

A process of change has begun which has a very long way to go. The arrival of the CPS as a powerful agency within the criminal justice system has focused a lot of minds. We may not have started the process, but our arrival has caused people to think about the whole of the criminal justice system in a different way, and much more quickly, than Philips or any of us would have imagined prior to 1986.[22] There is a whole range of issues where we can now bring influence to bear, like cautioning and diversion. It does not make sense to have the power to decide to discontinue or not to commence proceedings in the public interest unless you can also say: 'It is in the public interest for this potential defendant to be diverted to a particular agency.' On a small number of occasions, when we want to divert people away from prosecution to enable them to get help, we have no way of guaranteeing that this will actually happen. There are probation officers who are prepared to accept referrals, or it is possible to work through the defence solicitor; but the criminal justice system, if we are serious about diversion and cautioning, needs mechanisms and services whereby the public interest is served by not prosecuting these people. We are a very long way from that, because neither the framework nor the facilities exist to do that on a formal basis. The Crown Prosecutor's Code declares that it will not be right to prosecute a juvenile solely to secure access to the welfare powers of the court. In general terms, I believe the same principle must apply to adults also. However, there are occasions when the public interest may justify taking a person to court, since, unless we take proceedings, a person may not seek or be obliged to obtain the help they really need. It is in this area that I would like to see the powers of the CPS extended. The Code talks of the responsible use of our discretion at various stages of the prosecution process in the interest of the offender, the public, and of justice itself.

The theme of the widening scope of the CPS is pursued by B3, a chief crown prosecutor:

Court proceedings are in some cases futile, and there needs to be something apart from the court arena to dispose of certain categories of cases. The early drugs involvement of young teenagers provides an example, in that to punish them through the courts does not always achieve very much. Some of them are just experimenting, and once they have had their fling they will never become involved in crime again. Others are more in need of help and guidance, and a fine is merely going to increase the financial pressures. Here there is a need for something other than the court process, both to protect society and to meet the public interest. If the Crown Prosecution Service is going to be involved more generally in the administration of justice, rather than just prosecuting criminals, then it is essential that we are also 'involved' in the creation of schemes that are not court-based, even if it is only to the extent of knowing about their existence so that we can bear them in mind when making decisions in individual cases. We should help to set up 'cautioning-plus' schemes, and get other agencies to think very clearly about the sort of people that they are going to deal with, so that we can ensure that the public are protected.[23] Some other agencies can be so concerned with the needs of an individual that they forget the needs of the public.

We now have a central and a more 'political' position in the criminal justice system than we had under the old system. I would not describe myself as a political animal: I am essentially a lawyer who happens to be in the Crown Prosecution Service. One is, however, aware of the broad spectrum of the society within which we exist and the issues which are operating within that society. Should one, for example, be influenced in deciding whether to object to bail in an individual case by a knowledge that prisons are overcrowded? Especially when one knows that one can legitimately object to bail in the terms of the Bail Act, and that the magistrates are likely to grant your application for remand in custody? It is a dilemma particularly for me as a chief crown prosecutor, but it makes the role more challenging and stretches you as a human being. Certainly, I have moved a long way from what I envisaged I would be doing as a lawyer when I first qualified.

B4, a senior official with the Crown Prosecution Service, considers the CPS role with respect to sentencing, which had become an issue largely because of the powers introduced by the Criminal Justice Act 1988 for the CPS, through the Attorney General, to seek leave to appeal against 'unduly lenient' sentences:

The strong belief in the English legal system that the prosecutor should not be concerned with sentence is an utterly simplistic view because the way that the prosecution process deals with the defendant inevitably postulates a view about sentence, right down to the policeman who does not report the offence because he knows it will be a £2 fine. Provided it is done within a proper ambit, the prosecutor has scope for going for a charge that may carry a lesser penalty, or asking for summary trial rather than trial on indictment.

There is a very real concern that a prosecutor, in the sense of someone standing up in court, should be a minister of justice rather than a persecutor or an avenging angel; and there is a very strong resistance to the prosecutor actually asking for a particular sentence, as in some other jurisdictions, or seeking to give broad advice on appropriate sentences, which happens even in some jurisdictions with a more Anglo-Saxon tradition. There is still a lot of concern in some quarters that the CPS should not go in that direction. There are others who take a different view, and think that the tradition is not particularly well founded.

As suggested earlier, there is considerable scope for extending the notion of 'public interest' beyond the prosecution agency to other sections of criminal justice. This notion is explored with reference to sentencing by E6, a barrister and recorder, in the context of the sentencing proposals contained in the 1990 White Paper, *Crime, Justice and Protecting the Public*.[24]

I would like policy in relation to 17- to 20-year-olds to change to the extent that it is a rarity for a youngster to be remanded in custody. I would like to see the judiciary thinking not 'Can we avoid sending this person to prison?' but 'Do we have to send him to prison?' There is a tendency in Britain to start with custody. For many cases it seems to be the attitude of 'It's prison unless you can avoid it.' The Criminal Justice Act 1988 says that a young person aged 17 to 20 can only be sent into custody if certain conditions are fulfilled, and if the offence for which he is being sentenced is sufficiently serious. Judges sought to interpret that as meaning that the five or six cases put together that he might have pleaded guilty to would bring it into the category of being serious enough, and the Court of Appeal, in some cases reluctantly, has been forced to rule that the Act means what it says, and you cannot put all the things together that the lad pleaded to and say: 'This makes it serious enough for custody.' It has to be the one offence that is to be serious enough. That will gradually change attitudes.

I would like to see, for all defendants, of all ages, the attitude being: 'Can we best serve the community by passing a sentence that does not deprive this man of his liberty?' If you approach it that way, there are many things you can do. There is a tendency amongst the senior judiciary to confuse crime with sin. This is a terrible mistake. The public can be persuaded that community-based penalties are not the soft option and that they are better for the community. I would like to see almost no women sent to prison. There are enough reports to show that prison does have a different effect on them than it does on men. There will always be that element in the country who will say: 'We are here to punish and that is that.' I would like to see that become less prevalent. The White Paper's proposals, if properly put into practice, could be good, if we exclude silly things like electronic tagging, which was only put in as a sop to the right wing of the Tory party.[25]

For the probation service, the challenge is one of looking out beyond criminal justice so as to be at the forefront of crime prevention efforts and to explore fully the opportunities for the reintegration of offenders within the community. At the same time, there are pressures by central government that encourage probation officers to adopt more interventionist and controlling roles. The issue as set forth by C2, a chief probation officer, is whether the probation service will succumb to a more expedient posture or chart a way forward that builds upon its caring and supportive traditions:

> The probation service should be involved in diversion from crime, from the courts, and from custody. We are beginning to make an impact on custody. In terms of diversion from court, there are juvenile justice gatekeeping arrangements with the police and the creative roles we can forge with the Crown Prosecution Service. There is the whole unchartered territory of discontinuance. I would welcome probation staff being attached full-time to the CPS to provide social information which is of relevance to the decision whether to prosecute or not. One of the most challenging perspectives is diversion from crime. I would like to see more targeted intervention in estates where there is a record of crime, and where young people have a sense of being marginalized or alienated. That is a way ahead we can seek in conjunction with local authorities and the police service.
>
> The service has always occupied a strange position in the

criminal justice process. At its centre it deals with judges and magistrates, providing an information service for those at risk of custody. At its margins it is the gateway, or a social catalyst, which opens on to other opportunities for offenders, and which offers a more lasting opportunity for resettlement than the criminal justice process. There is that paradox of being at the centre and the margins, and mediating between the two.

We are moving from a public monopoly situation and entering a mixed economy market. The probation service has to grasp the challenge of partnerships with the private sector and with voluntary organizations. The stronger our alliances locally, the less likely we will be fragmented and broken up by central government. I am concerned that we do not become over-orthodox in terms of community-based penalties, and that we have a sense of the unorthodox.

Partnerships are very important. I need colleagues who are prepared to embrace those partnerships. We can no longer afford to be insular in our approaches. If we are to survive, we must develop less defensive postures, and that is the message one is giving to staff all the time. There is a possibility that you could devalue the work of probation officers, but they remain the key workers in terms of supervision in the probation order and community service. There is still a need for people with excellent communication skills, who are able to reach out in the community in imaginative kinds of ways and who believe in themselves and in the value of their contacts with people.

C5, a chief probation officer, emphasizes the need for the probation service to be clear about its philosophical base:

I was very strongly against proposals such as intermittent custody, which did not fit in at all with my belief that custody is only for those people that need to be there. If people can be let out during the week, then why bother getting them in at weekends? I believed that very strongly, and I had a sharp debate with colleagues who felt it was going to come, and so the probation service ought to be offering to do it. That was totally against my philosophy. I felt very strongly that we should say it is not right, and be quite clear about why it is not right, and I was absolutely delighted when the Home Office dropped their thinking about that. I have always felt the same about electronic surveillance. It does not fit in with the modern service, and the values which I still think are important. Also, if people can afford to be out of custody on an electronic tag, then they do not need to be in custody in the first place. It is about having a clear philosophical base, and being able to argue from that, and trying to be consistent with it. There is a danger around that the service becomes a bit expedient, and of thinking that if something is coming they should make the most of it.

C3, a chief probation officer, on why the probation service should hold its nerve and carefully define its responsibilities:

When I first saw some of the statements in the White and Green Papers on the future for the probation service, I found some of these perturbing in terms of just what was being asked of us. But I have been able to recognize that sometimes a megaphone was being used to address another audience. Sometimes statements were being made to be deliberately provocative to the probation service. To some extent, the crisis subsided for me when I realized it was a question of holding our nerve, improving those things that we knew it was important to improve on; not just playing the game, but being an organization that justified and deserved the position that is open to us in the 1990s as not just a provider of services, but also as a catalyst to ensure that other services exist that are supportive to the sort of work that we want to do. For a relatively small organization we are fairly influential, and the potential to be more influential is there, but in a different kind of way. I more than welcome the opportunity to move on from being the monopolistic service-provider. I would like to think that we are in a bridging position between the fulcrum of the criminal justice system and where the services need to be improved and delivered and built up, which is out in the communities where the vast majority of crime happens and has to be dealt with, recognizing that the courts handle only about 5 per cent of all of that.

It will be a testing experience, because we have got to be very careful about not just playing with (and appearing to own) the words that come so readily to the lips of the ministers when they are talking about criminal justice issues, and we must not be leaping on to bandwagons. There is still enormous potential to win and retain the confidence of and build our credibility with the courts, because we have got measures that address the balance of control and care, and that actually do deliver at the end of the day. The responsibility of organizations like the Association of Chief Officers of Probation and of individuals like me in the roles that we have is to point out that there is no quick-fix solution; there is not an easy option: you actually have to make an investment in working with people in conditions of liberty to enable them to use that liberty responsibly, and you cannot pander to the way in which it is usually discussed, which is about people just being deprived of liberty. With the exception of the relatively small proportion of people convicted of criminal offences from which all of us need to be protected and they from themselves, it is about learning how to use liberty.

The probation service has to be prepared to be much more explicit about the fact that it does have the authority of the courts behind it, that it does actually have a responsibility to address public concerns about crime, and should not just pretend that, with a little bit of help and support, tangible and material needs are met, and that is all that you need to do. It is much more about individuals challenging and confronting situations. You do have to point out that certain things will not be tolerated either by family or by friends or by the law. It is using that authority to the greatest extent possible; the fact that you are saying 'this matters', so it still does turn on the interaction between one person and another. They do not have to be friends, or even have to like each other very much, but they have to have some sort of respect for the position that each has. There has to be respect for the freedom of practitioners to do their job, and for the freedom of the supervisee to have the opportunity to learn to use liberty. We need to organize ourselves better and establish more shared goals. For a lot of main-grade staff, I am aware that it looks as though the management of the service has sold out to what the Home Office wants, and become ready to take on a more centralist role rather than a centre staff role for the service. There is, however, still great scope for professional judgement and discretion, and indeed at the end of the day I am absolutely relying on the good exercise of that professional discretion and judgement to handle issues of public protection, conflict, and change.

Tagging presents quite a crisis point for me. It goes against everything that experience has taught me about what enables offenders and their families to respond to the intervention of a criminal justice agency like our own. As a philosophy, it is weak and flawed. It panders to something that at the end of the day is just cosmetic. It looks like toughness and control, but it is just a lot of gimmickry that is destined to be a failure.

C1, another chief probation officer, on the dangers of a Credo Two pragmatism prevailing within the probation service:

I have no qualms that we are part of the criminal justice system. But our prime task is to be a social work service 'core', and that buys in a set of principles which we do not abandon because other bits of the criminal justice process find them uncomfortable. Some of the most furious debates recently have been whether my own organization, the Association of Chief Officers of Probation, is able to respond with one voice, because we are a fairly fragmented service. About a third of us are in one camp, and the other two-thirds take the much more pragmatic view that it is better to be in and influencing and going along with the government of the day so that the probation service retains a central position in the criminal justice sphere. I have some sympathy for that. But virtually all the agencies of the criminal justice process have got much more discretion than they ever give themselves credit for in the exercise of that discretion, from the police onwards.

There is an enormous amount of room for manœuvre, and I intend to exploit that room for manœuvre. The best example has been the national rules on the community service order. We fought quite a considerable battle to get them rewritten, to allow us more discretion, and I use that discretion quite substantially in running the community service scheme which meets our principles much more than it would if we literally followed the rules. But there are chief probation officers who would use their discretion to introduce tagging tomorrow if it was available. Tagging has no place in my area, and that may bring us into direct conflict with central government. Some of these arguments need to be considered coolly before the heat of the thing is on us.

The turn in attitudes among many practitioners, to a large extent, had to do with an increased awareness of the damaging effects of imprisonment. For a growing number of practitioners, the aim was not so much to reform the prison as to replace it. The driving vision was the de-escalation of the criminal justice process. One result was that questions about prison regimes and the organization of the prison system were neglected. The disturbances at Strangeways and other prisons in England and Wales in April 1990 and the subsequent wide-reaching inquiry by Lord Justice Woolf at least served in part to remedy the situation. The extracts that follow address three key issues that were to the fore in the late 1980s and in the wake of the disturbances of 1990: the connections between the prison system and other criminal justice agencies; the relationship of the prison and the general community; and the need to redefine the respective roles of staff and prisoners.

C4, a chief probation officer, regrets the gap between working with offenders in prison and in the community:

> In the past, an old stipendiary magistrate used to call probation officers the 'scavengers' of the court (picking up drunks and prostitutes). The post-war years saw us as 'servants' of the court. In the eighties we were partners in it all—an equal status and voice, along with judges, prosecutors, and the police. Some of them do not like it at present, but it is coming and I hope the nineties will see it come to fruition. The probation service has an incredible body of knowledge about how to deal with offenders, their circumstances, and their problems. That is why we aimed to inform the Woolf inquiry about what goes on in prison, but also what we do in the community. There is a great mismatch between the work with offenders in prison and the work in our day centres and hostels. They are on different planets, and that is crazy. Trying to refashion some of the work that goes on in prison over the next ten years would be very good news.

D5, a prison governor, on the need to locate the prison system within the wider criminal justice process:

> We must reinforce and confirm our links with the criminal justice system. Linked to that (for the next century, rather than

the next ten years) is a model which will give us a continuum between the community and the prison. Until such times as we pull all those strands together, we will never be able to deal with the offender as a person who came from and will return to the community. Underlying the whole concept of a prison is the notion of exile behind the high walls. Encouraging the community to accept responsibility for its offenders, or at least to have an involvement with them, is going to be one of the challenges. There must be much smaller community-based prisons. The problem is that we have this massive penal estate which bears no resemblance to this sort of concept.

We have to go back to first principles and recognize that imprisonment is about punishment and that punishment consists of the deprivation of liberty. It is important to have that statement recognized by the sentencer, so that no one is sent to prison for whom there is any reasonable alternative, and also that sentences should only be passed for the shortest necessary time, however long that may be. The two people who have always recognized that prison is essentially about the deprivation of liberty are the prisoner (when his cell door closes at night there is no handle on the inside) and the prison officer, who knows that his main task is making sure he locks up the same number of prisoners at night as he opened up in the morning. If one stops there, then prison becomes a very negative place, and those of us who started from that premiss were criticized for taking a very pessimistic and negative position— not least by one or two academics, who seemed to find that notion difficult to cope with.

Once we had established that premiss, we moved forward to say that, after the individual had passed the portals of the prison, one then took a different perspective of seeing how best one might make use of the time during which liberty has been taken away. The first priority of the prison service is custody, to ensure that the prisoners complete the sentence passed on them by the court. The vast majority of prisoners accept that they have to stay in prison until the sentence is finished. The second requirement is that prison should be a safe place for prisoners to live and for staff to work. The vast majority of prisoners will accept that requirement. The third level is that the system has an obligation to gather around itself as many resources and facilities as possible and to present them to the prisoner as opportunities to make use of his time in prison if he wishes to prepare himself for release, recognizing that ultimately the prisoner has the right to say no. Whether he does or does not should not be used as a measure of his fitness for release, which it was under the rehabilitative dimension, particularly within the context of parole. Such a stance makes it more difficult for the prison service to argue for more resources, on the grounds that people will be rehabilitated and at some point in the future the crime rate be reduced. Instead, you are asking for these resources so that prisons can develop positively and humanely. The reason for that is, in common humanity we have an obligation to prisoners, to people who are deprived of their liberty, to provide them with all these facilities. But it is a more honest position. Once you move down that road to say that we are encouraging prisoners to take up the opportunities which are presented, then you are beginning to recognize the prisoner as an individual.

Perhaps it is me who is being left behind and overtaken by the next generation, but I have difficulty coming to terms with the notion of the prison service as a business. I can understand the aspects in which it is a business and has to be run efficiently and effectively for a variety of reasons, but it may be my lack of perception in seeing it in terms of profit and loss accounts. Equally, seeing the prisoner as customer, which is a notion which is now being actively promoted, has a degree of artificiality about it which worries me slightly. I have no difficulty with the principle that the prisoner is a responsible individual with rights and has to be treated as such, to be consulted, and for there to be participation between him and the system. To move from that to the notion that the prisoner is a customer smacks a bit of artificiality, and of an arrogance which goes back to the rehabilitative ideal. The one thing the prisoner wants is to get out of prison. The one thing we cannot give him is release. So we say to him, 'I'm sorry, customer, release isn't on the agenda today. You can have everything else, but you can't have release.'

D6, a prison governor, on his unease about the dominant role of Credo Two values within the prison system:

I have become quite concerned about the management style of the prison service in the light of an increasing preoccupation with efficiency. The tail is beginning to wag the dog rather than the other way round. I am very committed to the philosophy of efficiency and effectiveness, because it is important that we get value for money out of the resources, and that management be in control and developing policies for the improvement of the treatment of prisoners. However, some aspects of the efficiency strategy have meant that the values of headquarters have been twisted. We no longer have the concern for prisoners at the forefront of our mind. Efficiency is at the forefront of our mind, more as an end than as a means, and that does concern me.

E3, a senior Home Office official, articulates the challenge to redefining roles and relationships within the prison:

The Prison Service has always had difficulty in achieving a properly adult relationship between management, staff, and prisoners. 'Adult' may not be quite the right word, but that triangular relationship has always been difficult.[26] There was a short period at the beginning of the 1980s when we tried to get some of these issues on the service's agenda; but I really never succeeded in communicating to the prison service what I meant when I talked about the 'justice model' as a shorthand description of what I was trying to say, and there were some people who found it deeply threatening to their own position and confidence. I was trying to ask about the basis of authority and respect in a prison setting—How do prisoners and staff talk to each other? Do they use their names? Do they think of each other as human beings with identities and personalities? What is the justification for uniform, and what kind of uniform should it be? Too often prisoners seemed somehow to be treated as the raw material of a process, with not very much attention to the human dynamics of life in what we know can sometimes become volatile institutions. We had just had the Wormwood Scrubs riot in 1979, which was a tragic and comprehensive failure of management and of the professionalism we would expect from the staff.[27] Of course, circumstances were previously very difficult, but once in my earlier days in the Prison Commission, a disturbance was taking place in the prison, and an assistant commissioner—Charles Cape—happened to be making a visit. Charles came through the door, the prisoners fell silent, and two or three of them said: 'Hello Sir, we didn't know you were in the prison.' They talked for a bit, and the prisoners went quickly back to their cells.[28] It was all more difficult by 1979, and perhaps is even more difficult now; but treating prisoners as people, on a basis of mutual respect, is part of running a stable institution, and I felt we had lost something over the intervening years.

D4, a prison governor, concludes this chapter with a radical challenge for the prison system:

> We have lost our way and must start to take seriously how we manage our prisons. We must search for a baseline or foundation on which to build. We must decide what is a successful prison. Apart from the need to hold people in secure custody, a successful prison is one where there is no fear, where human beings have dignity and a sense of worth, and where there is a well ordered community life. Any limitation imposed on an inmate's freedom of action should be the minimum necessary to ensure security and control. Prisoners are not second-class citizens and should not be treated as such.

The Practitioners displayed considerable skills in working within a myriad of constraints arising from the conventional structure of criminal justice agencies. Within these agencies, priority is given to narrowly defined performance measures and to short-term trouble-shooting over the articulation of general principles and objectives. The tendency is to sidestep basic deficiencies and injustices and to devise new ways of patching up existing arrangements. In short, the dominating Credo Two management style presents the most serious contemporary challenge to those practitioners who place liberal and humane values ahead of pragmatic expediency.

For the Credo Three practitioner to be effective, an unusual combination of reflection and action is crucial. This reflective action by the practitioner means that day-to-day practice is assessed against policy purposes, necessitating that questions are continually raised about what is going on. There must exist, furthermore, a recognition of the strategic role that practice is able to play in shaping policy. An important aspect is adeptness at building upon advantages that arise from making the connections across criminal justice. An especially effective vehicle for reform may be inter-agency collaboration around a specific task. Alliances by practitioners across agency boundaries, but also with persons and organizations outside the criminal justice arena, are often also crucial to successful reform.[29] Breaking down the barriers behind which criminal justice agencies operate is one preoccupation of the practitioner as policy-maker. The next and final chapter explores more widely this dynamic relationship between practitioners and the policy-making process.

# 7 Practice Leading Policy

THE preoccupation of criminal justice agencies is to get through the work at hand expediently. Wide horizens and long-term visions are eschewed in favour of moving matters along in accord with established procedures. This environment provides a comfortable workplace for the Credo Two practitioner, who is encouraged neither to consider the overall direction nor to be unduly distracted by casualties along the way. A common theme emerging from these narratives is the imperative to undermine this prevailing ethos. Adhering to a cluster of liberal and humane values, the Credo Three practitioner occupies an ambiguous and sometimes precarious position. The ambiguity arises because of the chasm that exists between the agency's mission statements, which often parade values and sentiments that go to the core of Credo Three, and routine practice. This institutionalized dissonance between words and deeds leaves the Credo Three practitioner in tune with formal purposes, but out of step with working traditions and culture.

The gap between the philosophy people say they espouse and what they actually do is highlighted by several Practitioners. As one of them (A1) observes, this can result in matters going terribly wrong, as was graphically exemplified by the series of miscarriages of justice in Britain which came to light in the late 1980s. A central task for the Royal Commission on Criminal Justice was to determine whether these events were aberrations in an otherwise sound process or symptomatic of a much deeper malaise. The instinctive Credo Two response downplays the severity of mistakes, suggesting instead that persons such as the Birmingham Six are innocent only in a technical sense; that errors are remarkably rare and are explicable in terms of unusual circumstances, such as grave public alarm in the aftermath of a terrorist outrage. They would suggest that remedial action should be confined to disciplining a handful of wayward practitioners and re-emphasizing formal objectives. Furthermore, the Credo Two reaction may also be to go on the offensive by inserting into the policy melting pot issues such as the right for suspects to remain silent and the usefulness of the jury. The fundamental transformation of the everyday ethos of the criminal justice workplace is thereby sidestepped.

For the Credo Three practitioner, these concerns prompt close attention to the ways in which words are used by agencies and, in particular, to language serving as a cosmetic shield that disguises what is being done. As Nils Christie has argued: 'Crime control has become a clean, hygienic operation. Pain and suffering have vanished from the text-books and from the applied labels.'[1] The Credo Three practitioner insists that words match deeds and that the meaning of these words is retained. This sceptical and detached stance is a crucial ingredient of the reflective action that characterizes the Credo Three practitioner.

In Britain during the 1980s, there were indications that this mode of reflective action was succeeding to some extent. There was a new sense of empowerment among Credo Three practitioners with respect to initiatives that might lead, shape, and, in due course, determine policy. Over this period, some unusually hopeful developments took place with reference to certain aspects of criminal justice. Especially notable were new approaches to young offenders.[2]

E1, a justices' clerk, illustrates how practitioners were able to redirect the juvenile justice policy agenda:

> When I started out, I was not in tune with government or official policy. In fact, there were some early confrontations in the campaign to get proper juvenile justice arrangements locally. This escalated until letters were being written to ministers. One result was a minister's reply to the effect that everything was contained in the Children and Young Persons Act 1969 and that new ideas would also be announced shortly. One of these turned out to be the 'short sharp shock' detention centre regime, which was soon to disappear, completely discredited. But by the time I became more involved at national level, my personal views seemed to coincide with the more liberal policies and attitudes which were now developing. By then, a lot of those in the magistracy and many justices' clerks were not happy with

the way things were. By the time of the Green Paper on 17- to 20-year-old young offenders, I could not help wondering if I had not written it myself![3] The ideas that were being expressed in the paper, I and others had been arguing about for the last five years.

E2, another a justices' clerk, describes how he resolved to de-escalate the sentencing practice of his court:

> It has taken a while to develop. The focus was sharpened by the background reading I did for my LLM [Master in Law] but had been shaped earlier when I had responsibility for training new magistrates and by the heightening publicity that we are an over-punitive country compared with the rest of western Europe; particularly by seeing people who come before the court and realizing that, by and large, what the court does makes no difference, that they are either going to grow out of it or there is going to be some fundamental shift such as getting married. Very few of the people I have seen coming before the courts are deterred by custody, and quite a few find life that much harder and therefore have more potential for crime afterwards.

There were some indications during the eighties that a turn in attitudes had taken place among many practitioners in favour of de-escalating the criminal justice process. In particular, there was a new awareness of the negative consequences that follow exposure to the formal criminal justice apparatus, and in particular to the courts and the prison system. Furthermore, practitioners appreciated that they did not have to wait for the emergence of a new policy framework or for some other form of lead from the centre, but might themselves be immediate catalysts of change. In this respect, practice appears to have both led and shaped policy.[4] Finally, practitioners recognized that reform strategies often rest upon collaboration that extends across agency boundaries. A keener awareness of the interdependence of criminal justice underpinned the rich variety of schemes involving the partnership of two or more agencies.

It is important not to overstate the extent of this turn in practice. Account must also be taken of several countervailing tendencies resulting in a confusing and contradictory collage of policy and practice. For example, during the eighties there was a marked increase in prison sentence lengths with respect to both the time imposed by the courts and the period actually served.[5] Greater severity for some offenders coinciding with reduced severity for others was presented by Home Office ministers as explicit expression of the 'twin-track' idea which they espoused. This 'twin-track' notion, which found some expression in the Criminal Justice Act 1991, promises to be escalatory in its net effects. Demanding more punishment for certain categories of offenders feeds the culture of severity. Furthermore, in practice, the line between serious and less serious offenders may be blurred and applied arbitrarily.[6]

There is another problem which arises from the Criminal Justice Act 1991 and relates directly to issues raised in this book. Much of the progress achieved during the eighties was the result of initiatives taken by practitioners at the local level. The Act sought a consistent approach across the country, especially with respect to sentencing practice. The legislation, along with other developments such as new fiscal controls, gave priority to central over local initiatives for criminal justice reform.[7] Aspects of this issue are addressed by several Practitioners in Chapter 5, who agree on the imperative to retain considerable scope for manœuvre at the local level. If that were lost, the attractions of the job for them would largely disappear. Despite the new central controls, it should not be beyond the ingenuity of practitioners to identify and exploit the opportunities which remain at the local level.

The de-escalatory turns of practice, however, have not been without impact upon policy. For example, in a robust acknowledgement of the negative consequences of custody, the Home Office observed that, 'however much prison staff try to inject a positive purpose into the regime, as they do, prison is a society that requires virtually no sense of personal responsibility from prisoners. Normal social or working habits do not fit.'[8] For good measure, the authors of the White Paper went on to say: 'It is unrealistic to expect most prisoners to emerge at the end of their sentence as reformed characters; imprisonment provides many opportunities to learn criminal skills from other inmates. Custody can have a devastating effect on some prisoners and on their families.'[9] The degree of change in the official outlook on imprisonment can be gauged by the Home Office stance during the 1980s to statutory restrictions on the use of custodial sentences. In 1982 the government opposed, unsuccessfully as it turned out, the introduction of criteria to be satisfied before a person under the age of 21 could be sentenced to custody. In 1988, however, the government accepted amendments that tightened these statutory criteria. Three years later, in the Criminal Justice Act 1991, the government itself took the initiative to extend this approach to all custodial sentencing.

This nascent and tentative de-escalatory movement in Britain can be instructively compared with developments in the Netherlands some thirty years earlier. The 'Utrecht School' is an especially potent example of Credo Three on an ascendant path. Based at the Institute of Criminology at the University of Utrecht, a handful of academics not only made a profound impact upon the working ideology of a generation of lawyers, but also were influential across the criminal justice process and within central government. As David Downes has put it:

Their appeal was philosophical: a form of existentialist and phenomenological reinstatement of the offender as a human being with individual rights, against which the penal system was seen as imposing an authoritarian and often unjust, as well as counter-productive, and unnecessarily oppressive, regime.[10]

Willem de Haan's summary of the Utrecht School's message encapsulates the essence of Credo Three. He suggests that the school reflected a

strong empathy with the delinquent as a fellow human being. Central to their thinking was the notion that the convict is, on the one hand, a person needing help and, on the other hand, entitled to certain basic rights. In other words, compassion, co-responsibility and a deep sense of humanity supplied the main motives for the School's critique of institutions and conditions which do not do justice to the delinquent's basic rights.[11]

Between 1950 and 1975, an increasing proportion of cases were disposed of by public prosecutors rather than by the courts. With respect to sentencing, Dutch judges greatly reduced prison sentence lengths.

More recently, between 1983 and 1990, prison numbers fell by about 20 per cent in the Federal Republic of Germany. As

Johannes Feest has argued, this decline was not related to recorded crime trends or to socio-economic indicators.[12] Once again, the decisive factor appears to have been a substantial shift in the working ideologies of practitioners. With reference to the sharp drop in the number of young offenders in German prisons, John Graham concluded that: '... it looks as though judges have fundamentally altered the basis on which they sentenced young offenders. Prison is increasingly being used as a final resort and is avoided wherever possible. The overall approach to young offenders is one of patience until they grow out of crime.'[13]

However, examples of Credo Three taking root and having a substantial impact remain few and far between. Developments in the United States, for example, have been in the opposite direction. During the 1980s, the US prison and jail population more than doubled to reach a total of 1.2 million persons. A culture of severity across the United States appears to have stimulated the values and sentiments of Credo One among many practitioners.[14] In Britain and other parts of Europe the picture is less clear, but the Credo Three practitioner at least has a fragile presence within the criminal justice process, thereby offering some relief to a scene that is mostly bleak and chilling.

The urgent contemporary requirement remains to radically transform working ideologies at all stages of the criminal justice process. The components of this task were spelled out by Jerome Skolnick in the early 1970s with respect to the police and are amended here to address a broader canvas.

Skolnick believed that the needed philosophy of practitioners must rest upon a set of values conveying the idea that criminal justice agencies are as much institutions dedicated to the achievement of legality and humanity in society as they are official social organizations designed to control misconduct through the invocation of punitive sanctions. What must occur, he felt, is a significant alteration in working ideologies, so that these rest on the values of a democratic legal order, rather than on technological proficiency or pragmatic expediency.[15]

For practitioners, the challenge is to extend their reflective action to all aspects of the criminal justice apparatus, and to challenge its inclination towards tidy consensus.[16] In Britain during the 1990s, some encouragement for these endeavours may be generated by the aspirations of the Woolf report on prisons, by the work of the Royal Commission on Criminal Justice, and through closer contacts with countries across Europe. But of most significance will be the practitioners themselves. It is with them that hope must reside if, at least for a while, criminal justice is to be made a little more decent.

# NOTES

## Preface

1. Steven Marcus, 'Their Brothers' Keepers: An Episode from English History', in William Gaylin (ed.), *Doing Good: The Limits of Benevolence* (New York, Pantheon, 1978), 42.

## Chapter 1: Working Credos

1. See David Garland, *Punishment and Modern Society* (Oxford University Press, 1990).

2. Winston S. Churchill speaking on the Prisons Vote, HC *Debates*, 5th Series, vol. 19, col. 1354, 20 July 1910. Churchill's speech is one of the truly classic statements on the limits of criminal justice in a free society. On Churchill's brief but remarkable period as Home Secretary, see Andrew Rutherford, 'Lessons from a Reductionist Era', in Philippe Robert and Clive Emsley (eds.), *History and Sociology of Crime* (Pfaffenweiler, Centaurus-Verlagsgesellschaft, 1990), 58–63.

3. Jerome Miller, 'The Flight from Meaning: Convicts and their Keepers in the 21st Century', 6th Annual Lecture to the Institute of Criminal Justice, University of Southampton (March 1991). Jerry Miller is one of the most remarkable criminal justice practitioners of his generation. He stumbled into this arena almost by mistake in 1969 when, to his considerable surprise, he was appointed commissioner of the Massachusetts Department of Youth Services. His successful efforts to transform that agency from managing custody to providing a rich array of community-based services has been thoroughly documented by Lloyd Ohlin and colleagues at Harvard Law School; see especially R. B. Coates, A. D. Miller, and L. E. Ohlin, *Diversity in a Youth Correctional System: Handling Delinquents in Massachusetts* (Cambridge, Mass., Ballinger, 1978). After going on to other state appointments in Illinois and Pennsylvania, he founded, in 1979, the National Center on Institutions and Alternatives. The Center provides defence counsel with 'creative sentencing' packages and is also an independent voice on corrections in the United States. For Miller's own account of the Massachusetts story, see Jerome Miller, *Last One Over the Wall* (Columbus, Ohio State University Press, 1991). For Miller, 'it was [also] a human passage—raucous, fitful, threatening, exhilarating, at times impulsive, often unpredictable, changing direction as we took advantage of unexpected opportunities—and always

difficult. We lived for a time on the edge of bureaucracy, professional ethics, legality, and politics. But our small deinstitutionalization challenged many of the preconceptions that sustain most state institutions and provide the rationale for the mishandling of many of the mentally ill, the homeless, the delinquent, and the criminal.' (ibid, p. xiii)

4. *The Independent* (13 April 1991).

5. Walter B. Miller, 'Ideology and Criminal Justice Policy: Some Current Issues', *Journal of Criminal Law and Criminology*, 64 (1973), 141-62 at 142.

6. See e.g. A. M. Colman and P. L. Gorman, 'Conservatism, Dogmatism and Authoritarianism in Police Officers', *Sociology*, 16 (1982), 1-11; J. S. Carroll, W. T. Perkowitz, A. J. Lurigio, and F. M. Weaver, 'Sentencing Goals, Causal Attributions and Personality', *Journal of Personality and Behaviour*, 52 (1987), 107-18; K. B. Melvin, L. K. Grambling, and W. M. Gardner, 'A Scale to Measure Attitudes toward Prisoners', *Criminal Justice and Behaviour*, 12 (1985), 241-53; Nigel G. Fielding and Jane Fielding, 'Police Attitudes to Crime and Punishment', *British Journal of Criminology*, 31 (1991), 39-53. An important recent addition is Robert Reiner's study of chief constables. Reiner interviewed 40 of the 43 chief constables in England and Wales in 1986-7: see Robert Reiner, 'Thinking at the Top: A Survey of Chief Constables' Attitudes and Opinions', *Policing*, 5 (1989), 181-99; and Robert Reiner, *Chief Constables: Bobbies, Bosses, or Bureaucrats?* (Oxford University Press, 1991). Extending beyond criminal justice, and in the tradition of Studs Terkel, *Working People Talk About What They Do All Day and How They Feel About What They Do* (New York, Pantheon, 1974), is Sholom Glouberman, *Keepers: Inside Stories from Total Institutions* (London, King Edward's Hospital Fund for London, 1990). For a wide-ranging exploration of police attitudes, see Roger Graef, *Talking Blues: The Police in their Own Words* (London, Fontana, 1990).

7. Stanton Wheeler, Edna Bonacish, M. Richard Cramer, and Irving K. Zola, 'Agents of Delinquency Control: A Comparative Analysis', in Stanton Wheeler (ed.), *Controlling Delinquents*, New York, John Wiley (1968), 31-60 at 55.

8. Ibid. 57.

9. John Hogarth, *Sentencing as a Human Process* (University of Toronto Press, 1971), 337.

10. Francis A. Allen, *The Crimes of Politics: Political Dimensions of Criminal Justice* (Cambridge, Mass., Harvard University Press, 1974), 22.

11. Jerome H. Skolnick, *Justice Without Trial: Law Enforcement in a Democratic Society*, 2nd edn. (New York, John Wiley, 1975), 196.

12. Ibid. 61.

13. Herbert L. Packer, *The Limits of the Criminal Sanction* (Stanford University Press, 1968); see also Herbert L. Packer, 'The Models of the Criminal Process', *University of Pennsylvania Law Journal*, 113 (1964), 1—68.

14. The term 'hands off' was first used to characterize the traditional stance of the courts to prisons in the USA in Note, 'Beyond the Ken of the Courts: A Critique of Judicial Refusal to Review the Complaints of Convicts', *Yale Law Journal*, 72 (1963), 506—58.

15. Doreen McBarnett, 'False Dichotomies in Criminal Justice Research', in John Baldwin and A. Keith Bottomley (eds.), *Criminal Justice: Selected Readings* (Oxford, Martin Robertson, 1978), 23—34, at 30.

16. Michael McConville and John Baldwin, *Courts, Prosecution, and Conviction* (Oxford University Press, 1981), 188.

17. Ibid. 211—22.

18. A. E. Bottoms and J. D. McClean, *Defendants in the Criminal Process* (London, Routledge & Kegan Paul 1976), 233.

19. Robert Reiner, *The Politics of the Police* (Brighton, Wheatsheaf, 1985), 103.

20. David Garland, op. cit. n. 1, p. 183.

21. Ibid. 184.

22. 'Practitioner' is capitalized when reference is made to the interviewees.

23. In 1981 Lord Lane, the Lord Chief Justice, refused to allow a study of judicial attitudes to sentencing to proceed. Lord Lane stated that research into the attitudes, beliefs, and reasoning of judges was not the way to obtain an accurate picture: sentencing was an art and not a science, and the further judges were pressed to articulate their reasons, the less realistic the exercise would become: Andrew Ashworth *et al.*, *Sentencing in the Crown Court: Report of an Exploratory Study*, Centre for Criminological Research, University of Oxford, Occasional Paper no. 10 (1984), 64. In the event, twenty-five judges were interviewed.

24. There have been several recent important studies of criminal justice élites. David Downes provided a portrait of senior criminal justice policy-makers and practitioners in The Netherlands: see his *Contrasts in Tolerance: Post-War Penal Policy in The Netherlands and England and Wales* (Oxford University Press, 1988). Paul Rock has charted the dynamics of criminal justice reform with reference to victims' issues in both Canada and England and Wales: see his

*A View From The Shadows: The Ministry of the Solicitor General of Canada and the Justice for Victims of Crime Initiatives* (Oxford University Press, 1986); and *Helping Victims of Crime: The Home Office and the Rise of Victim Support in England and Wales* (Oxford University Press, 1990). Robert Reiner's study of chief constables in England and Wales is a particularly valuable contribution: *Chief Constables: Bobbies, Bosses, or Bureaucrats?* op. cit. n. 6.

25. The one exception had retired from his position of chief constable a few years earlier.

26. Christian Pfeiffer has argued that many German judges and prosecutors who were part of the contemporary movement away from the use of prisons are 'the children of the sixties': Christian Pfeiffer, 'A European Perspective', Conference on 'Young Offenders: A Chance to Get it Right', Jersey Probation Service (November 1990).

27. On the 'new consciousness', see Charles A. Reich, *The Greening of America* (Harmondsworth, Penguin, 1970).

28. Jerome Bruner, *Acts of Meaning* (Cambridge, Mass., Harvard University Press, 1990), 123. Bruner's interest is to produce 'spontaneous autobiographies' by 'listening to people in the act of creating a longitudinal version of self' (p. 120).

29. Downes, op. cit., n. 24, p. 85.

30. Michael Billig *et al.*, *Ideological Dilemmas: A Social Psychology of Everyday Thinking* (London, Sage, 1988), 2.

31. Donald Polkingthorne, *Narrative Knowing and the Human Sciences* (Albany: SUNY Press, 1988), 150; cited by Bruner, op. cit. n. 27 at pp. 115–16.

32. The letters A, B, C, D, and E refer, respectively, to the police, the Crown Prosecution Service, the probation service, the prison service, and a miscellaneous category that includes persons working within the courts and a Home Office civil servant.

33. Miller, op. cit. n. 5, p. 142.

34. Packer (1968), op. cit. n. 13, p. 154.

35. James Fitzjames Stephen, *A History of the Criminal Law of England* (London, Macmillan, 1883), 81–2. For an extreme expression of the logic of Credo One, see Ingo Muller, *Hitler's Justice: The Courts of the Third Reich* (London, Tauris, 1991), esp. 68–81, on the enthusiastic contributions of many legal academics to Nazi jurisprudence. The authoritarian model of criminal justice is perceptively analysed by Leon Radzinowicz, 'Penal Regressions', *Cambridge Law Journal*, 50(3) (1991), 422–44.

36. Vulnerable prisoner units hold persons (often sexual offenders) who

request protection under Rule 43 (Prison Rules, England and Wales). 'Nonce' is prison parlance for sex offender. For a useful overview, see Prison Reform Trust, *Sex Offenders in Prison* (London, Prison Reform Trust, 1990).

37. James B. Jacobs, *Stateville: The Penitentiary in Mass Society* (University of Chicago Press, 1977), 103—4.

38. Lord Denning actually said: 'If the six men fail, it will mean that much time and money and worry will have been expended by many people for no good purpose. If the six men win, it will mean that the police were guilty of perjury, that they were guilty of violence and threats, that the confessions were involuntary and were improperly admitted in evidence and that the convictions were erroneous. That would mean the Home Secretary would either have to recommend they be pardoned or he would have to remit the case to the Court of Appeal ... This is such an appalling vista that every sensible person in the land would say: "It cannot be right these actions should go any further."' *Mcllkenny* v *Chief Constable of West Midlands Police Force and Another* [1980] 2 All ER 227, at 239—40. An insightful analysis of this type of posture is provided by Mike McConville, Andrew Sanders, and Roger Leng, *The Case for the Prosecution, Police Suspects and the Construction of Criminality* (London, Routledge, 1991), 173—90.

39. For a useful historical overview of the liberal position on crime and criminal justice, see Leon Radzinowicz, *Ideology and Crime: A Study of Crime in its Social and Historical Context* (London, Heinemann Educational, 1966), 1—28.

40. See David Garland, 'Critical Reflections', in Huw Rees and Eryl Hall Williams, *Punishment, Custody and the Community: Reflections and Comments on the Green Paper* (London School of Economics, 1989), 15—16.

41. Jerome Miller, cited in Andrew Rutherford, *Growing Out of Crime* (London, Penguin, 1986), 104.

42. Jerome Miller, *Last One Over the Wall* (Columbus, Ohio State University Press, 1991), xiii.

43. Martha Minow, 'A Tribute to Justice Thurgood Marshall', *Harvard Law Review*, 105(1) (1991), 66—76, at 70—2.

44. William J. Brennan, 'A Tribute to Justice Thurgood Marshall', *Harvard Law Review*, 105(1) (1991), 23—33, at 32—3.

45. Louk Hulsman of Erasmus University, Rotterdam, has been especially forceful in his analysis of the damage caused by the apparatus of criminal justice: see Louk Hulsman, 'Critical Criminology and the Concept of Crime', *Contemporary Crises*, 10 (1986), 63—80; and 'Penal Reform in The Netherlands. Part 1:

Bringing the Criminal Justice System Under Control', *Howard Journal*, 20 (1981), 150–9.

46. Sir Richard Mayne, the founding co-commissioner of the Metropolitan Police, wrote that the function of the police was: 'The prevention of crime . . . the protection of life and property, the preservation of public tranquillity': *New Police of the Metropolis* (1829).

47. Michael Jenkins wrote: 'My starting point is therefore that the prison system has a tendency to create more problems than it receives and has an equal tendency to fail inmates because, out of its survival fear, it tends to respond to corporate threats, real or imaginary, rather than the real problems of inmates': 'Control Problems in Dispersals', in Anthony E. Bottoms and Roy Light, *Problems of Long-Term Imprisonment* (Aldershot, Gower, 1987), 261–80, at 262.

## Chapter 2: The Practice Context

1. For a fuller historical assessment, see Terence Morris, *Crime and Criminal Justice since 1945* (Oxford, Basil Blackwell, 1989).

2. A. E. Bottoms, 'An Introduction to "The Coming Crisis" ', in A. E. Bottoms and R. H. Preston (eds.), *The Coming Penal Crisis* (Edinburgh, Scottish Academic Press, 1980), 2.

3. Nils Christie, *Limits to Pain* (Oxford, Martin Robertson, 1982), 47.

4. See e.g. James Q. Wilson, *Thinking About Crime*, (New York, Vintage Books, 1975), and John Dilulio, *Governing Prisons: A Comparative Study of Correctional Management* (New York, Free Press, 1987).

5. Christie, op. cit. n. 3 at 49.

6. Robert Martinson never actually used the phrase 'nothing works', although it did become his epitaph. His conclusion in 1974 was: 'With few and isolated exceptions, the rehabilitative efforts that have been reported so far have had no appreciable effect on recidivism' (Martinson, 'What Works? Questions and Answers about Prison Reform', *The Public Interest* (Spring 1974), 22–54, at 25. Later, in 1979, he wrote: 'Contrary to my previous position, some treatment programs do have an appreciable effect on recidivism . . . New evidence from our current study leads me to reject my original conclusion': 'New Findings, New Views: A Note of Caution Regarding Sentencing Reform', *Hofstra Law Review*, 7 (1979), 243–58 at 252.

7. See, e.g., American Friends Service Committee, *Struggle for Justice* (New York, Hill and Wang, 1971).

8. David Garland, *Punishment and Modern Society* (Oxford University Press, 1990), 186.

9. R. D. Laing's remarks were made on a television programme, *Did You Used To Be R. D. Laing?* broadcast on Channel 4 in Britain (April 1991).

10. David Garland, 'Critical Reflections', in Huw Rees and Eryl Hall Williams (eds.), *Punishment, Custody and the Community: Reflections and Comments on the Green Paper* (London School of Economics, 1989), 4–18 at 8–9.

11. *Penal Practice in a Changing Society*, Cmd. 645 (London, HMSO, 1959).

12. *Report of the Interdepartmental Committee on the Business of the Criminal Courts* (Chairman, Mr Justice Streatfeild), Cmnd. 1289 (London, HMSO, 1961).

13. Hermann Mannheim and Leslie T. Wilkins, *Prediction Methods in Relation to Borstal Training* (London, HMSO, 1955).

14. Alexander Paterson hand-picked a remarkable group of men to be Borstal housemasters. He was a member of the Prison Commission of England and Wales from 1922 to 1946. See, generally, S. K. Ruck (ed.), *Paterson on Prisons* (London, Frederick Muller, 1951).

15. S. R. Brody, *The Effectiveness of Sentencing: A Review of the Literature*, Home Office Research Study no. 35 (London, HMSO, 1976).

16. Paul Gendreau and R. R. Ross, 'Revivification of Rehabilitation: Evidence from the 1980s', *Justice Quarterly*, 4 (1987): 349–408; see also Francis T. Cullen and Paul Gendreau, 'The Effectiveness of Correctional Rehabilitation: Reconsidering the "Nothing Works" Debate', in Lynne Goodstein and Doris Layton MacKenzie (eds.), *The American Prison: Issues in Research and Policy* (New York, Plenum Press, 1989), 23–44.

17. *Final Report of the Royal Commission on the Police* (Chairman, Sir Henry Willink), Cmnd. 1728 (London, HMSO, 1962). For overviews of the subsequent trend towards more centralized control of the police, see Laurence Lustgarten, *The Governance of Police* (London, Sweet and Maxwell, 1986), and Robert Reiner, *Chief Constables: Bobbies, Bosses, or Bureaucrats* (Oxford University Press, 1991), esp. 249–300.

18. Lord Scarman, *The Brixton Disorders*, Cmnd. 8427 (London, HMSO, 1981). Lord Scarman has been described as 'the most splendid and notorious liberal among the law lords': Noel Annan, *Our Age: The Generation that Made Post-War Britain* (London, Fontana, 1991), 416. Leslie Scarman would have been a fascinating

addition to this study. He exercised a profound humanizing influence across a wide range of policy issues during his judicial career from 1961 to 1986. After his retirement, he was actively involved in the campaigns on behalf of the Guildford Four and regarding other miscarriages of justice. For his work as a judge, see Simon Lee, *Judging Judges* (London, Faber & Faber, 1988), esp. 154–62. See also the profile on his eightieth birthday in the *Guardian*, 29 July 1991.

19. For an absorbing account of the consequences of the long-drawn-out miners' dispute upon attitudes to the police, see Penny Green, *The Enemy Without: Policing and Class Consciousness in the Miners' Strike* (Milton Keynes, Open University Press, 1990).

20. *The Confait Case: Report by the Hon. Sir Henry Fisher*, HC 90 (London, HMSO, 1971); *Report of the Royal Commission on Criminal Procedure*, Cmnd. 8092 (London, HMSO, 1981); the resulting statutes were the Police and Criminal Evidence Act 1984 and the Prosecution of Offences Act 1985.

21. See, in particular, Robert Kee, *Trial and Error: The Maguires, the Guildford Pub Bombings and British Justice* (London, Hamish Hamilton, 1986); Grant McKee and Ros Franey, *Time Bomb: Irish Bombers, English Justice and the Guildford Four* (London, Bloomsbury, 1988); and Chris Mullin, *Error of Judgement: The Birmingham Bombings* (London, Chatto & Windus, 1986).

22. The setting up of the Royal Commission on Criminal Justice (under the chairmanship of Lord Runciman) was announced after the release of the Birmingham Six in March 1991.

23. The powers to stop and search suspected persons were codified by the Police and Criminal Evidence Act 1984.

24. See e.g. Tony Vinson, *Wilful Obstruction: The Frustration of Prison Reform* (Melbourne, Methuen, 1982). Jerry Miller has also described how he attempted to transform the traditional young offender institutions in Massachusetts into therapeutic communities; for every step forward, the institutions later appeared to slide back two paces. See Jerome Miller, *Last One Over the Wall* (Columbus, Ohio State University Press, 1991). In some circumstances, the reformer may be able to use the eventual backlash to positive advantage. For an account of oscillatory processes within Henderson Hospital during the period Maxwell Jones pioneered the therapeutic community approach to 'psychopathic' patients, see Robert N. Rapoport, *Community as Doctor: New Perspectives on a Therapeutic Community* (London, Tavistock, 1960), esp. 135–42.

25. George Blake had served five years of a forty-two-year sentence for offences under the Official Secrets Act; see *The Report of the Inquiry*

*into Prison Escapes and Security* (undertaken by Lord Mountbatten), Cmnd. 3175, (London, HMSO, 1966).

26. See generally Andrew Rutherford, *Prisons and the Process of Justice* (Oxford University Press, 1986).

27. See *Prison Disturbances, April 1990*, Report of an Inquiry by Lord Justice Woolf and Judge Stephen Tumim, Cm. 1456 (London, HMSO, 1991).

28. In 1983 the Secretary of State for Scotland followed the decision taken in England and Wales and restricted parole for certain categories of prisoners with sentences of over five years.

29. For a pioneering study of the transformation of the environment that shaped penal policy and practice over 1895−1914, see David Garland, *Punishment and Welfare: A History of Penal Strategies* (Aldershot, Gower, 1985).

## Chapter 3: Patterns of Early Development

1. John Hogarth, *Sentencing as a Human Process* (University of Toronto Press, 1971), 342.

2. Richard Hoggart, *The Uses of Literacy: Aspects of Working-Class Life with Special Reference to Publications and Entertainments* (London, Chatto & Windus, 1957); E. P. Thompson, *The Making of the English Working Class* (London, Gollancz, 1963); Raymond Williams, *Culture and Society, 1780−1950* (London, Chatto and Windus, 1958).

3. See e.g. Kenyon J. Scudder, *Prisoners are People* (New York, Doubleday, 1952).

4. For an appreciation of David Hewlings and his work in the Prison Service, see the obituary in the *Independent* (7 January 1991), which included the following remarks: 'Those who had the privilege of working with him found the experience inspiring, rewarding and deeply fulfilling, for he knew that it was only by gaining the wholehearted commitment of staff through the involvement with each other and with prisoners that prisons could become instructive communities with the potential for growth and development. But he also knew that prisons should not be isolated from the wider society. He was alive to the contribution that other organizations could make and to the need to work in partnership with them, whether in industry, the health and education services, or in the voluntary sector.'

5. Erving Goffman, *Asylums: Essays on the Social Situation of Mental Patients and Other Inmates* (Garden City, NY, *Doubleday*, 1961).

6. On Michael Jenkins see also Ch. 1, n. 47.

7. Dr Norman Jepson was Professor of Adult Education at the University of Leeds and for many years taught at the Prison Service Staff College.

8. Richard Titmuss and David Donnison were members of the Department of Social Administration at the London School of Economics (sometimes referred to as the 'Department of Applied Virtue').

9. Howard Becker, *Outsiders: Studies in the Sociology of Deviance* (New York, Free Press, 1963); Howard Parker, *View from the Boys* (Newton Abbot, David and Charles, 1974).

10. Goffman, *Asylums*, op. cit. n. 5; *The Presentation of Self in Everyday Life* (Garden City, NY, Doubleday, 1959).

11. Person serving a sentence of preventive detention. This sentence (which was abolished in 1958) was used to imprison repeat offenders for longer periods than would otherwise be justified.

12. For an insightful account of the relatively narrow focus of the work of the CID compared with other police officers, see Dick Hobbs, *Doing the Business: Entrepreneurship, the Working Class, and Detectives in the East End of London* (Oxford University Press, 1988), esp. 197–216.

13. On contacts within like-minded professional colleagues, see also pp. 111–16.

## Chapter 4: Turning-Points

1. Serpico was a New York City detective who exposed corruption among his colleagues.

2. Attendance at conferences was not mentioned as a turning-point by any of the Practitioners. This contrasts with a study of the victims movement in Canada, in which Paul Rock noted the particular impact that conferences on victimology had had, in terms of being a *rite de passage* or *a crise de conscience*, for some of the key personnel: Paul Rock, *A View From The Shadows: The Ministry of the Solicitor General of Canada and the Justice for Victims of Crime Initiative* (Oxford University Press, 1986), 101–4.

3. Philip G. Zimbardo *et al.*, *Influencing Attitudes and Changing Behaviour: An Introduction to Theory and Applications of Social Control and Personal Power* (New York, Random House, 1977); and Stanley Milgram, *Obedience to Authority: An Experimental View* (London, Tavistock, 1974).

4. See also pp. 111–12 below.

5. Radical Alternatives to Prison (RAP) was founded in 1970; the

National Association for the Care and Resettlement of Offenders (NACRO) was established in 1966.

6. Report of the Advisory Council on the Penal System, *Non-custodial and Semi-custodial Penalties* (chaired by Baroness Wootton) (London, HMSO, 1970).

7. Thomas Mathiesen, *Prison on Trial: A Critical Assessment* (London, Sage, 1990), 137.

8. Ibid. 140.

9. For a description of the Home Office's 'inner circle' of élites, see Paul Rock, *Helping Victims of Crime: The Home Office and the Rise of Victim Support in England and Wales* (Oxford University Press, 1990), 31–6.

## Chapter 5: Individuals and Agencies

1. A Belfast solicitor, killed by a 'loyalist' paramilitary organization in February 1989.

2. The local branch of the Prison Officers' Association.

3. See e.g. Norman Tutt, 'A Decade of Policy', *British Journal of Criminology*, 21 (1981), 246–56.

4. David Thomas, *Principles of Sentencing*, 2nd edn. (London, Heinemann, 1979).

5. The 'instant caution' is intended to avoid the delays incurred by consulting social services and other agencies which are associated with the traditional process of deciding upon the formal caution as an alternative to prosecution.

6. This very important insight illuminates one of the dynamics of the criminal justice process, especially in its early stages. See, generally, Malcolm M. Feeley, *The Process is the Punishment: Handling Cases in a Lower Criminal Court* (New York, Russell Sage Foundation, 1979).

7. *Opportunity and Responsibility: Developing New Approaches to the Management of the Long-Term Prison System in Scotland*, Scottish Prison Service (Edinburgh, HMSO, 1990).

8. Advisory Council on the Penal System, *The Regime for Long-term Prisoners in Conditions of Maximum Security* (Chairman, Professor Leon Radzinowicz) (London, HMSO, 1968). The Radzinowicz Report argued that top security prisoners be dispersed across a number of prisons where they should be held in relatively relaxed conditions. The government preferred this option to Lord Mountbatten's recommendation in 1966 that such prisoners be concentrated in one or two super-maximum-security prisons.

9. On A3's self-education, see p. 75.

10. Lord Scarman carried out the inquiry into the Brixton disturbances in 1981: *Brixton Disorders, 10–12 April 1981: Report of an Inquiry Presented to Parliament by the Secretary of State for the Home Department*, Cmnd. 8427 (London, HMSO, 1981). His recommendation that there be police consultative groups was enacted in the Police and Criminal Evidence Act 1984.

11. See Robert Reiner, *Chief Constables: Bobbies, Bosses, or Bureaucrats?* (Oxford University Press, 1991), esp. 110–17.

12. See, in particular, *Punishment, Custody, and the Community*, Cm. 424 (HMSO, London, 1988); *Crime, Punishment and Protecting the Public*, Cm. 965 (London, HMSO, 1990); *Supervision and Punishment in the Community: A Framework for Action*, Cm. 966 (London, HMSO, 1990); *Organising Supervision and Punishment in the Community: A Decision Document* (London, Home Office, 1991).

13. See David Garland, 'Critical Reflections', in Huw Rees and Eryl Hall Williams (eds.), *Punishment, Custody and the Community: Reflections and Comments on the Green Paper* (London School of Economics, 1989), 4–18.

14. Home Office, *Statement of National Objectives and Priorities* (London, Home Office, 1984).

15. See e.g. John McCarthy's letter to *The Times* on leaving the prison service (19 November 1981). McCarthy wrote: 'I did not join the Prison Service to manage overcrowded cattle-pens; nor did I join to run a prison where the interests of the individuals have to be sacrificed continually to the interests of the institution; nor did I join to be a member of a service where staff that I admire are forced to run a society that debases.' He ended his letter by writing that he could not for much longer tolerate 'the inhumanity of the system within which I work'.

## Chapter 6: Ways Forward

1. Surprisingly, there is no standard text on criminal justice reform. For a select bibliography, see Roger Hood, 'Criminology and Penal Change: A Case Study of the Nature and Impact of Some Recent Advice to Governments', in Roger Hood (ed.), *Crime, Criminology and Public Policy* (London, Heinemann, 1974, 375–417); Gordon Rose, *The Struggle for Penal Reform* (London, Stevens, 1961); for a comprehensive portrayal of criminal justice reform in England and Wales between 1880 and 1914, see Leon Radzinowicz and Roger Hood, *A History of English Criminal Law*, V, *The Emergence of Penal Policy* (London, Stevens, 1986). The classic American historical studies are David Rothman, *The Discovery of the Asylum:*

*Social Order and Disorder in the New Republic* (Boston, Little Brown, 1971) and *Conscience and Convenience: The Asylum and its Alternatives in Progressive America* (Boston, Little Brown, 1980). Essential additions to this literature are Paul Rock's comparative studies of the policy-making process with respect to victims of crime: see Paul Rock, *A View from the Shadows: The Ministry of the Solicitor General of Canada and the Justice for Victims of Crime Initiative* (Oxford University Press, 1986) and *Helping Victims of Crime: The Home Office and the Rise of Victim Support in England and Wales* (Oxford University Press, 1990). For a seminal American study, see Lloyd E. Ohlin, 'Conflicting Interests in Correctional Objectives', in Richard A. Cloward *et al.*, *Theoretical Studies in Social Organisation of the Prison* (New York, Social Science Research Council, Pamphlet 15, 1960), 111–29. See also Alden D. Miller and Lloyd E. Ohlin, *Delinquency and Community: Creating Opportunities and Controls* (London, Sage, 1985), esp. 73–102. For another key American study of interest groups, see Richard A. Berk and Peter Rossi, *Prison Reform and State Elites* (Cambridge, Mass., Ballinger, 1970).

2. Sir Alec Clegg, who was chief education officer for West Riding County Council in 1945–74.

3. See in particular *The Challenge of Crime in a Free Society: A Report by the President's Commission on Law Enforcement and Administration of Justice* (Washington, DC, US Government Printing Office, 1967), 7–12; Malcolm M. Feeley and Austin D. Sarat, *The Policy Dilemma: The Crisis of Theory and Practice in the Law Enforcement Assistance Administration* (Minneapolis, University of Minnesota Press, 1980); and Alfred Blumstein, 'Coherence, Co-ordination and Integration in the Administration of Criminal Justice', in Jan van Dijk *et al.* (eds.), *Criminal Law in Action: An Overview of Current Issues in Western Societies* (Deventer, Kluwer, 1988), 247–58.

4. Rock (1990), op. cit. n. 1, p. 39.

5. See the evidence of the police professional organizations to the House of Commons Home Affairs Committee's inquiry into *The Crown Prosecution Service*, Session 1989/90, HC 118–11 (London, HMSO, 1990). For the views of several chief constables on the Crown Prosecution Service, see Robert Reiner, *Chief Constables: Bobbies, Bosses, or Bureaucrats?* (Oxford University Press, 1991), 156–60.

6. This phrase is attributed to the Dutch practitioner–scholar, Jan van Dijk.

7. *Code of Practice*, Crown Prosecution Service (London, HMSO, 1986); see, generally, Andrew Ashworth, 'The "Public Interest" Element in Prosecution', *Criminal Law Review* (1987): 595–607.

8. See Christopher Stone, *Bail Information for the Crown Prosecution Service* (New York, Vera Institute of Justice, 1988).

9. See Christopher Stone, *Public Interest Case Assessment*, (New York, Vera Institute of Justice, 1989). A full account of Stone's remarkable impact in these two areas of British criminal justice is awaited. His approach is very much in the Vera Institute tradition of 'an immense self-confidence that has allowed it to maintain a truly experimental outlook, admit shortcomings, and rethink solutions... The more significant innovation is the innovation-producing institution itself': Malcolm M. Feeley, *Court Reform on Trial* (New York, Basic Books, 1983), 219.

10. See e.g. Arthur Rossett and Donald R. Cressey, *Justice by Consent* (Philadelphia, Lippincott, 1976); A. S. Blumberg, 'The Practice of Law as a Confidence Game: Organizational Cooption of a Profession', *Law and Society Review*, 1 (1967), 15-39; Michael McConville and John Baldwin, *Courts, Prosecution, and Conviction* (Oxford University Press, 1981). Contrasting 'administrative' and 'rights' reform strategies, Malcolm Feeley has written: 'Perhaps the greatest danger with the administrative strategy is that it will work, and that it will transform a contentious and embattled group of professionals into co-operative bureaucrats': op. cit. n. 9, p. 205.

11. This danger also exists wherever special collaborative arrangements are set up by criminal justice agencies to deal with particular offences or offending situations. Some examples of how this can happen, when magistrates' courts have to deal with large numbers of defendants charged with public order offences, are given in Andrew Rutherford and Bryan Gibson, 'Special Hearings', *Criminal Law Review* (July 1987): 440-8.

12. The time limits on detaining subjects as set forth in the *Codes of Practice for the Detention, Treatment and Questioning of Persons by the Police* under the Police and Criminal Evidence Act 1984.

13. See e.g. Roger Tarling, *Sentencing Practice in Magistrates' Courts*, Home Office Research Study no. 56, (London, HMSO, 1989), 28.

14. In October 1990, after this interview had taken place, Lord Lane, the Lord Chief Justice, by way of a Practice Direction, issued national mode of trial guidelines which apply to all practitioners across the criminal justice process.

15. David Moxon, *Sentencing in the Crown Courts*, Home Office Research and Planning Unit, Report no. 103 (London, HMSO, 1989), esp. 57-9.

16. See the Woolf Report, *Prison Disturbances April 1990*, Report of an Inquiry by Lord Justice Woolf and Judge Stephen Tumim, Cm. 1456 (London, HMSO, 1991), esp. 260-4.

17. Home Office, *Custody, Care and Justice: The Way Ahead for the Prison Service in England and Wales*, Cm. 1647 (London, HMSO, 1991), 10.

18. These powers are provided by the Criminal Justice Act 1991.

19. See *Report of the Committee on Remuneration and Conditions of Service of Certain Grades in the Prison Services* (Chairman, Mr Justice Wynn-Parry), Home Office and Scottish Home Department, Cmnd. 544 (London and Edinburgh, HMSO, 1958). By contrast, Lord Justice Woolf found the prison service to be inward-looking, and he urged that it recognize and accept a more central role in relation to the criminal justice process: see Woolf Report, op. cit. n. 16, pp. 260–4.

20. This is a reference to the various powers of the police to stop and search suspected persons that existed prior to the Police and Criminal Evidence Act 1984. One survey found that 'West Indians' were four times as likely to be stopped as 'white people' when on foot and that 'West Indians' were much more likely to be stopped repeatedly: David J. Smith, *Police and People in London*, i, *A Survey of Londoners* (London, Policy Studies Institute, 1983), 94–100.

21. These guidelines were set out in Home Office Circular 14/85 and were superseded by Home Office Circular 59/90, which further strengthened the presumption in favour of cautioning young offenders. The respective responsibilities of the police and CPS were not, however, altered.

22. Sir Cyril Philips was chairman of the Royal Commission on Criminal Procedure. The *Report*, Cmnd. 8092 (London, HMSO, 1981), led to the Police and Criminal Evidence Act 1984 and to the Prosecution of Offences Act 1985.

23. 'Cautioning plus' refers to schemes that go beyond formally cautioning the offender by making a referral to a counselling programme or other specialist services. These schemes do raise fundamental problems. The absence of due process protections for unconvicted persons and the dangers of widening the criminal justice net are the main objections to formalizing arrangements of this sort. See Home Office Circular 59/90, which discouraged the setting up of 'cautioning plus' schemes.

24. Home Office, *Crime, Justice and Protecting the Public*, Cm. 965 (London, HMSO, 1990).

25. Electronic tagging was first used in Britain, on an experimental basis, in 1988 with respect to persons as a condition of their bail. Despite a generally unfavourable assessment by Home Office researchers, the Criminal Justice Act 1991 contains powers extending electronic surveillance to the sentencing stage: see George Mair and Claire Nee,

*Electronic Monitoring Unit Report: The Trials and Their Results*, Home Office Research Study no. 120 (London, HMSO, 1990).

26. On reviewing the text of his interview, E3 added: 'It is about what Woolf calls justice in prison': see the Woolf Report, op. cit. n. 16, esp. pp. 244–5.

27. See *Statement on the Background, Circumstances and Action Taken Subsequently relative to the Disturbance in 'D' Wing at HM Prison, Wormwood Scrubs, on 31 August 1979; together with the Report of an Enquiry by the Regional Director of the South East Region of the Prison Department*, HC 198 (London, HMSO, 1982).

28. Charles Cape ('of Good Hope', as he sometimes referred to himself) joined the prison service during the 'Paterson era' and retired in the mid-1960s.

29. Paul Rock has provided an intriguing case study of Irwin Waller, a 'moral entrepreneur' who consummately displayed these skills. 'Waller constructed social networks, disseminated ideas, and seized lines of action as they developed, forcing them together': Paul Rock, *A View from the Shadows: The Ministry of the Solicitor General of Canada and the Justice for Victims of Crime Initiative* (Oxford University Press, 1986), 380. Waller, as the director general of research in the Ministry of the Solicitor General, 'believed that political action could flow out of research. He engaged in tireless campaigning. He sustained and drew sustenance from an intellectual community of those who were bent on making policy' (p. 118).

## Chapter 7: Practice Leading Policy

1. Nils Christie, *Limits to Pain* (Oxford, Martin Robertson, 1982), 16; see also the glossary of 'controltalk' in Stanley Cohen, *Visions of Social Control: Crime, Punishment and Classification* (Cambridge, Polity Press, 1985), 273–81. Unsurpassed as a general guide is Orwell's 'principle of newspeak' in George Orwell, *1984* (New York, Harcourt Brace Jovanovich, 1949), 246–56.

2. During 1982–9, the number of juveniles receiving custodial sentences in England and Wales fell from 7400 to 2400: *Criminal Statistics, England and Wales, 1989*, Cm. 1322 (London, HMSO, 1990), 168. See, generally, Rob Allen, 'Out of Jail: The Reduction in the Use of Penal Custody for Male Juveniles, 1981–88', *Howard Journal of Criminal Justice*, 30 (1991), 30–52; and Howard Parker, Maggie Sumner, and Graham Jarvis, *Unmasking the Magistrates: The 'Custody or Not' Decision in Sentencing Young Offenders* (Milton Keynes, Open University Press, 1989).

3. *Punishment, Custody and the Community*, Cm. 424 (London, HMSO, 1988).

4. See Andrew Rutherford, 'The Mood and Temper of Penal Policy: Curious Happenings in England and Wales during the 1980s', *Youth and Policy*, 27 (1989), 27–31.

5. For males aged 21 and over, average sentence lengths imposed in the crown courts rose from 17.2 to 20.5 months over 1980–9: *Criminal Statistics, England and Wales, 1989*, Cm. 1322 (London, HMSO, 1990), 171.

6. Stanley Cohen has argued that bifurcation (i.e. the 'twin-track' strategy) has the pragmatic purpose of unclogging criminal justice by diverting minor cases so that resources can be 'rationally concentrated on the real business of crime control': Cohen, op. cit., n. 1, p. 128.

7. On the traditionally reactive stance of the Home Office, see Paul Rock, *Helping Victims of Crime: The Home Office and the Rise of Victims Support in England and Wales* (Oxford University Press, 1990), esp. 29–39.

8. *Crime, Justice and Protecting the Public*, Cm. 965 (London, HMSO, 1990), 6.

9. Ibid. 11. David Garland has drawn attention to how practitioners are taken into account in the drafting of policy documents, in terms of both practical consequences and symbolic significance: see his *Punishment and Modern Society* (Oxford University Press, 1990), 262–3. In this instance, the authors of the White Paper may have sought to reassure practitioners that the Home Office fully recognized the extent to which an anti-custody ethos had gained ground over the previous decade.

10. David Downes, *Contrasts in Tolerance: Post-War Penal Policy in the Netherlands and England and Wales* (Oxford University Press, 1988), 94.

11. Willem de Haan, *The Politics of Redress: Crime, Punishment and Penal Abolition* (London, Unwin Hyman, 1990), 69.

12. Johannes Feest, *Reducing the Prison Population: Lessons from the West German Experience?* (London, National Association for the Care and Resettlement of Offenders, 1988).

13. John Graham, 'Decarceration in the Federal Republic of Germany', *British Journal of Criminology*, 30 (1990), 150–70, at 162; see also Andrew Rutherford, 'The English Penal Crisis: Paradox and Possibilities', in R. Rideout and J. Jowell (eds.), *Current Legal Problems 1988* (London, Stevens, 1988), 93–113, esp. 107–9.

14. See, generally, Franklin E. Zimring and Gordon Hawkins, *The Scale of Imprisonment* (University of Chicago Press, 1991). See also Joan Mullen, 'State Responses to Prison Crowding: The Politics of Change', in Stephen D. Gottfredson and Sean McConville (eds.),

*America's Correctional Crisis: Prison Populations and Public Policy* (London, Greenwood Press, 1988), 79–109. Mullen argues that criminal justice élites in the United States often have opportunities to win support for reform efforts: 'If the public thought of criminal sanctioning as a practice with no articulatable goal, no rational procedure, and no authoritative leader, it might be less tolerant of perpetual crisis appropriations and more supportive of formal structures charged with bringing rationality to our chaotic corrections enterprise' (pp. 107–9).

15. Skolnick wrote: 'The needed philosophy of professionalism must rest on a set of values conveying the idea that the police are as much an institution dedicated to the achievement of legality in society as they are an official social organisation designed to control misconduct through the invocation of punitive sanctions . . . What must occur is a significant alteration in the ideology of police, so that police "professionalism" rests on the values of a democratic legal order, rather than on technological proficiency': Jerome H. Skolnick, *Justice Without Trial: Law Enforcement in Democratic Society*, 2nd edn. (New York, John Wiley, 1975), 238–9.

16. See especially, Mike McConville, Andrew Sanders, and Roger Leng, *The Case for the Prosecution, Police Suspects and the Construction of Criminality* (London, Routledge, 1991), 202–8.

# BIBLIOGRAPHY

Allen, Francis A., *The Crimes of Politics: Political Dimensions of Criminal Justice* (Cambridge, Mass., Harvard University Press, 1974).

Allen, Rob, 'Out of Jail: The Reduction in the Use of Penal Custody for Male Juveniles, 1981–88', *Howard Journal of Criminal Justice*, 30 (1991): 30–52.

American Friends Service Committee, *Struggle for Justice* (New York, Hill and Wang, 1971).

Annan, Noel, *Our Age: The Generation that Made Post-War Britain* (London, Fontana, 1991).

Ashworth, Andrew, 'The "Public Interest" Element in Prosecution', *Criminal Law Review* (1987): 595–607.

—— 'Criminal Justice and the Criminal Process', *British Journal of Criminology*, 28 (1988): 111–23.

—— Genders, Elaine, Mansfield, Graham, Peay, Jill, and Player, Elaine, *Sentencing in the Crown Court: Report of an Exploratory Study*, Centre for Criminological Research, University of Oxford, Occasional Paper no. 10 (1984).

Becker, Howard, *Outsiders: Studies in the Sociology of Deviance* (New York, Free Press, 1963).

Berk, Richard A., and Rossi, Peter, *Prison Reform and State Elites* (Cambridge, Mass., Ballinger, 1970).

Billig, Michael, *et al.*, *Ideological Dilemmas: A Social Psychology of Everyday Thinking* (London, Sage, 1988).

Blumberg, A. S., 'The Practice of Law as a Confidence Game: Organizational Cooption of a Profession', *Law and Society Review*, 1 (1967): 15–39.

Blumstein, Alfred, 'Coherence, Co-ordination and Integration in the Administration of Criminal Justice', in Jan van Dijk *et al.* (eds.), *Criminal Law in Action: An Overview of Current Issues in Western Societies* (Deventer, Kluwer, 1988), 247–58.

Bottoms, A. E., 'An Introduction to "The Coming Crisis"', in A. E. Bottoms and R. H. Preston (eds.), *The Coming Penal Crisis* (Edinburgh, Scottish Academic Press, 1980).

—— and McClean, J. D., *Defendants in the Criminal Process* (London, Routledge, 1976).

Brennan, William J., 'A Tribute to Justice Thurgood Marshall', *Harvard Law Review*, 105 (1), (1991): 23–33, at 32–3.

Brody, S. R., *The Effectiveness of Sentencing: A Review of the Literature*, Home Office Research Study no. 35 (London, HMSO, 1976).

Bruner, Jerome, *Acts of Meaning* (Cambridge, Mass., Harvard University Press, 1990).

Carroll, J. S., Perkowitz, W. T., Lurigio, A. J., and Weaver, F. M., 'Sentencing Goals, Causal Attributions and Personality', *Journal of Personality and Behaviour*, 52 (1987): 107–18.

Christie, Nils, *Limits to Pain* (Oxford, Martin Robertson, 1982).

Coates, R. B., Miller, A. D., and Ohlin, L. E., *Diversity in a Youth Correctional System: Handling Delinquents in Massachusetts* (Cambridge, Mass., Ballinger, 1978).

Cohen, Stanley, *Visions of Social Control: Crime, Punishment and Classification* (Oxford, Polity Press, 1985).

Colman, A. M., and Gorman, P. L., 'Conservatism, Dogmatism and Authoritarianism in Police Officers', *Sociology*, 16 (1982): 1–11.

Crown Prosecution Service, *Code of Practice* (London, HMSO, 1986).

Cullen, Francis T., and Gendreau, Paul, 'The Effectiveness of Correctional Rehabilitation: Reconsidering the "Nothing Works" Debate', in Lynne Goodstein and Doris Layton MacKenzie (eds.), *The American Prison: Issues in Research and Policy* (New York, Plenum Press, 1989), 23–44.

de Haan, Willem, *The Politics of Redress: Crime, Punishment and Penal Abolition* (London, Unwin Hyman, 1990).

Dilulio, John J., *Governing Prisons: A Comparative Study of Correctional Management* (New York, Free Press, 1987).

Downes, David, *Contrasts in Tolerance: Post-War Penal Policy in The Netherlands and England and Wales* (Oxford University Press, 1988).

Feeley, Malcolm, M., *The Process is the Punishment: Handling Cases in a Lower Criminal Court* (New York, Russell Sage Foundation, 1979).

—— *Court Reform on Trial* (New York, Basic Books, 1983).

—— and Sarat, Austin D., *The Policy Dilemma: The Crisis of Theory and Practice in the Law Enforcement Assistance Administration* (Minneapolis, University of Minnesota Press, 1980).

Feest, Johannes, *Reducing the Prison Population: Lessons from the West German Experience?* (London, National Association for the Care and Resettlement of Offenders, 1988).

Fielding, Nigel G., and Fielding, Jane, 'Police Attitudes to Crime and Punishment', *British Journal of Criminology*, 31 (1991): 39–53.

Garland, David, *Punishment and Welfare: A History of Penal Strategies* (Aldershot, Gower, 1985).

—— 'Critical Reflections', in Huw Rees and Eryl Hall Williams (eds.), *Punishment, Custody and the Community: Reflections and Comments on the Green Paper* (London School of Economics, 1989), 4–18.

—— *Punishment and Modern Society* (Oxford University Press, 1990).

Gendreau, Paul, and Ross, R. R., 'Revivification of Rehabilitation: Evidence from the 1980s', *Justice Quarterly*, 4 (1987): 349–408.

Glouberman, Sholom, *Keepers: Inside Stories from Total Institutions* (London, King Edward's Hospital Fund for London, 1990).

Goffman, Erving, *The Presentation of Self in Everyday Life* (Garden City, NY, Doubleday, 1959).

—— *Asylums: Essays on the Social Situation of Mental Patients and Other Inmates* (Garden City, NY, Doubleday, 1961).

Graef, Roger, *Talking Blues: The Police in their Own Words* (London, Fontana, 1990).

Graham, John, 'Decarceration in the Federal Republic of Germany', *British Journal of Criminology*, 30 (1990): 150–70.

Green, Penny, *The Enemy Without: Policing and Class Consciousness in the Miners' Strike* (Milton Keynes, Open University Press, 1990).

Hobbs, Dick, *Doing the Business: Entrepreneurship, the Working Class, and Detectives in the East End of London* (Oxford University Press, 1988).

Hogarth, John, *Sentencing as a Human Process* (University of Toronto Press, 1971).

Hoggart, Richard, *The Uses of Literacy: Aspects of Working Class Life with Special Reference to Publications and Entertainments* (London, Chatto and Windus, 1957).

Home Affairs Committee, Fourth Report, *The Crown Prosecution Service*, Session 1989/90, HC 118–1 and 11 (1990).

Home Office, *Penal Practice in a Changing Society*, Cmnd. 645 (London, HMSO, 1959).

—— *Report of the Interdepartmental Committee on the Business of the Criminal Courts* (chairman, Mr Justice Streatfeild), Cmnd. 1289 (London, HMSO, 1961).

—— *The Report of the Inquiry into Prison Escapes and Security* (undertaken by Lord Mountbatten), Cmnd. 3175 (London, HMSO, 1966).

—— Report of the Advisory Council on the Penal System: *The Regime for Long-term Prisoners in Conditions of Maximum Security* (chaired by Professor Leon Radzinowicz) (London, HMSO, 1968).

—— Report of the Advisory Council on the Penal System: *Non-custodial and Semi-custodial Penalties* (chaired by Baroness Wootton) (London, HMSO, 1970).

—— *The Confait Case: Report by the Hon. Sir Henry Fisher*, HC 90 (London, HMSO, 1971).

—— *Statement on the Background, Circumstances and Action taken subsequently relative to the Disturbance in 'D' Wing at HM Prison, Wormwood Scrubs on 31 August 1979; together with the Report of an Enquiry by the Regional Director of the South East Region of the Prison Department*, HC 198 (London, HMSO, 1982).

—— *Statement of National Objectives and Priorities* (statement on the probation service) (London, Home Office, 1984).

—— *Punishment, Custody and the Community*, Cm. 424 (London, HMSO, 1988).

Home Office, *Crime, Justice and Protecting the Public*, Cm. 965 (London, HMSO, 1990).

—— *Supervision and Punishment in the Community: A Framework for Action*, Cm. 966 (London, HMSO, 1990).

—— *Criminal Statistics, England and Wales, 1989*, Cm. 1322 (London, HMSO, 1990).

—— *Organising Supervision and Punishment in the Community: A Decision Document* (London, Home Office, 1991).

—— *Custody, Care and Justice: The Way Ahead for the Prison Service in England and Wales*, Cm. 1647 (London, HMSO, 1991).

—— and Scottish Home Department, *Report of the Committee on Remuneration and Conditions of Service of Certain Grades in the Prison Services* (chairman, Mr Justice Wynn-Parry) (London and Edinburgh, HMSO, 1958).

Hood, Roger, 'Criminology and Penal Change: A Case Study of the Nature and Impact of Some Recent Advice to Governments', in Roger Hood (ed.), *Crime, Criminology and Public Policy* (London, Heinemann, 1974), 375–417.

Hulsman, Louk, 'Penal Reform in The Netherlands. Part 1: Bringing the Criminal Justice System Under Control', *Howard Journal*, 20 (1981): 150–9.

—— 'Critical Criminology and the Concept of Crime', *Contemporary Crises*, 10 (1986): 63–80.

Jacobs, James B., *Stateville: The Penitentiary in Mass Society* (University of Chicago Press, 1977).

Jenkins, Michael, 'Control Problems in Dispersals', in Anthony E. Bottoms and Roy Light, *Problems of Long-Term Imprisonment* (Aldershot, Gower, 1987), 261–80.

Kee, Robert, *Trial and Error: The Maguires, the Guildford Pub Bombings and British Justice* (London, Hamish Hamilton, 1986).

Lee, Simon, *Judging Judges* (London, Faber & Faber, 1988).

Lustgarten, Laurence, *The Governance of Police* (London, Sweet & Maxwell, 1986).

Mair, George, and Nee, Claire, *Electronic Monitoring: The Trials and Their Results*, Home Office Research Study, no. 120 (London, HMSO, 1990).

Mannheim, Hermann, and Wilkins, Leslie T., *Prediction Methods in Relation to Borstal Training* (London, HMSO, 1955).

Marcus, Steven, 'Their Brothers' Keepers: An Episode from English History', in William Gaylin (ed.), *Doing Good: The Limits of Benevolence* (New York, Pantheon, 1978).

Martinson, Robert, 'What Works? Questions and Answers about Prison Reform', *The Public Interest* (Spring 1974): 22–54.

—— 'New Findings, New Views: A Note of Caution Regarding Sentencing Reform', *Hofstra Law Review*, 7 (1979): 243–58.

Mathiesen, Thomas, *Prison On Trial: A Critical Assessment* (London, Sage, 1990).

McBarnett, Doreen, 'False Dichotomies in Criminal Justice Research', in John Baldwin and A. Keith Bottomley (eds.), *Criminal Justice: Selected Readings* (Oxford, Martin Robertson, 1978), 23–34.

McConville, Michael, and Baldwin, John, *Courts, Prosecution, and Conviction* (Oxford University Press, 1981).

——, Sanders, Andrew and Leng, Roger, *The Case for the Prosecution, Police Suspects and the Construction of Criminality* (London, Routledge, 1991).

McKee, Grant, and Franey, Ros, *Time Bomb: Irish Bombers, English Justice and the Guildford Four* (London, Bloomsbury, 1988).

Melvin, K. B., Grambling, L. K. and Gardner, W. M., 'A Scale to Measure Attitudes toward Prisoners', *Criminal Justice and Behaviour*, 12 (1985): 241–53.

Milgram, Stanley, *Obedience to Authority: An Experimental View* (London, Tavistock, 1974).

Miller, Alden D., and Ohlin, Lloyd. E., *Delinquency and Community: Creating Opportunities and Controls* (London, Sage, 1985).

Miller, Jerome, 'The Flight from Meaning: Convicts and their Keepers in the 21st Century', Sixth Annual Lecture, Institute of Criminal Justice, University of Southampton (March 1991).

—— *Last One Over the Wall* (Columbus, Ohio State University Press, 1991).

Miller, Walter B., 'Ideology and Criminal Justice Policy: Some Current Issues', *Journal of Criminal Law and Criminology*, 64 (1973): 141–62.

Minow, Martha, 'A Tribute to Justice: Thurgood Marshall', *Harvard Law Review*, 105 (1) (1991): 66–76, at 70–2.

Morris, Terence, *Crime and Criminal Justice since 1945* (Oxford, Basil Blackwell, 1989).

Moxon, David, *Sentencing in the Crown Courts*, Home Office Research and Planning Unit, Report no. 103 (London, HMSO, 1989).

Mullen, Joan, 'State Responses to Prison Crowding: The Politics of Change', in Stephen D. Gottfredson and Sean McConville (eds.), *America's Correctional Crisis: Prison Populations and Public Policy* (London, Greenwood Press, 1988), 79–109.

Muller, Ingo, *Hitler's Justice: The Courts of the Third Reich* (London, Tauris, 1991).

Mullin, Chris, *Error of Judgement: The Birmingham Bombings* (London, Chatto and Windus, 1986).

Note, 'Beyond the Ken of the Courts: A Critique of Judicial Refusal to Review the Complaints of Convicts', *Yale Law Journal*, 72 (1963): 506–58.

Ohlin, Lloyd, E., 'Conflicting Interests in Correctional Objectives', in Richard E. Cloward *et al.*, *Theoretical Studies in Social Organisation of the Prison* (New York, Social Science Research Council, Pamphlet 15, 1960), 111–29.

Orwell, George, *1984* (New York, Harcourt Brace Jovanovich, 1949).

Packer, Herbert L., 'The Models of the Criminal Process, *University of Pennsylvania Law Journal*, 113 (1964): 1–68.

—— *The Limits of the Criminal Sanction* (Stanford University Press, 1968).

Parker, Howard, *View from the Boys* (Newton Abbot, David and Charles, 1974).

—— Sumner, Maggie, and Jarvis, Graham, *Unmasking the Magistrates: The 'Custody or Not' Decision in Sentencing Young Offenders* (Milton Keynes, Open University Press, 1989).

Pfeiffer, Christian, 'A European Perspective', unpublished paper presented to the Conference on 'Young Offenders: A Chance to Get it Right', Jersey Probation Service (November 1990).

Polkingthorne, Donald, *Narrative Knowing and the Human Sciences* (Albany, NY, SUNY Press, 1988).

Prison Reform Trust, *Sex Offenders in Prison* (London, Prison Reform Trust, 1990).

Radzinowicz, Leon, *Ideology and Crime: A Study of Crime and its Social and Historical Context* (London, Heinemann Educational, 1966).

—— 'Penal Regressions', *Cambridge Law Journal*, 50 (3) (1991), 422–44.

—— and Hood, Roger, *A History of English Criminal Law*, v, *The Emergence of Penal Policy* (London, Stevens, 1986).

Rapoport, Robert N., *Community as Doctor: New Perspectives on a Therapeutic Community* (London, Tavistock, 1960).

Reich, Charles A., *The Greening of America* (Harmondsworth, Penguin, 1970).

Reiner, Robert, *The Politics of the Police* (Brighton, Wheatsheaf, 1985).

—— 'Thinking at the Top: A Survey of Chief Constables' Attitudes and Opinions', *Policing*, 5 (1989): 181–99.

—— *Chief Constables: Bobbies, Bosses or Bureaucrats?* (Oxford University Press, 1991).

Rock, Paul, *A View From The Shadows: The Ministry of the Solicitor General of Canada and the Justice for Victims of Crime Initiative* (Oxford University Press, 1986).

—— *Helping Victims of Crime: The Home Office and the Rise of Victim Support in England and Wales* (Oxford University Press, 1990).

Rose, Gordon, *The Struggle for Penal Reform* (London, Stevens, 1961).

Rossett, Arthur, and Cressey, Donald R., *Justice by Consent* (Philadelphia, Lippincott, 1976).

Rothman, David, *The Discovery of the Asylum: Social Order and Disorder in the New Republic* (Boston, Little Brown, 1971).

—— *Conscience and Convenience: The Asylum and its Alternatives in Progressive America* (Boston, Little Brown, 1980).

Royal Commission on Criminal Procedure *Report* (chairman, Sir Cyril Philips), Cmnd. 8092 (London, HMSO, 1981).

Royal Commission on the Police, *Final Report* (chairman, Sir Henry Willinck), Cmnd. 1728 (London, HMSO, 1962).

Ruck, S. K. (ed.), *Paterson on Prisons* (London, Frederick Muller, 1951).

Rutherford, Andrew, 'A Statute Backfires: The Escalation of Youth Incarceration in England During the 1980s', in Jameson Doig (ed.), *Criminal Corrections: Ideals and Realities* (Lexington, Mass., Lexington Press, 1983), 73–91.

—— *Growing Out of Crime: Society and Young People in Trouble* (London, Penguin, 1986).

—— *Prisons and the Process of Justice* (Oxford University Press, 1986).

—— 'The English Penal Crisis: Paradox and Possibilities', in R. Rideout and J. Jowell (eds.), *Current Legal Problems 1988* (London, Stevens, 1988), 93–113.

——'The Mood and Temper of Penal Policy: Curious Happenings in England and Wales during the 1980s', *Youth and Policy*, 27 (1989): 27–31.

—— 'Lessons from a Reductionist Era', in Philippe Robert and Clive Emsley (eds.), *History and Sociology of Crime* (Pfaffenweiler, Centaurus-Verlagsgesellschaft, 1990), 58–63.

—— and Gibson, Bryan, 'Special Hearings', *Criminal Law Review* (July 1987): 440–8.

Scarman, Lord, *Brixton Disorders, 10–12 April 1981: Report of an Inquiry Presented to Parliament by the Secretary of State for the Home Department*, Cmnd. 8427 (London, HMSO, 1981).

Scottish Office, *Opportunity and Responsibility: Developing New Approaches to the Management of the Long-Term Prison System in Scotland*, Scottish Prison Service (Edinburgh, HMSO, 1990).

Scudder Kenyon J., *Prisoners are People* (New York, Doubleday, 1952).

Skolnick, Jerome H., *Justice Without Trial: Law Enforcement in a Democratic Society*, 2nd edn. (New York, John Wiley, 1975).

Smith, David J., *Police and People in London*, i, *A Survey of Londoners* (London, Policy Studies Institute, 1983).

Stephen, James Fitzjames, *A History of the Criminal Law of England*, 3 vols. (London, Macmillan, 1883).

Stone, Christopher, *Bail Information for the Crown Prosecution Service* (New York: Vera Institute of Justice, 1988).

—— *Public Interest Case Assessment* (New York, Vera Institute of Justice, 1989).

Tarling, Roger, *Sentencing Practice in Magistrates' Courts*, Home Office Research Study no. 56 (London, HMSO, 1989).

Terkel, Studs, *Working People Talk About What They Do All Day and How They Feel About What They Do* (New York, Pantheon, 1974).

Thomas, David, *Principles of Sentencing*, 2nd edn. (London, Heinemann, 1979).

Thompson, E. P., *The Making of the English Working Class* (London, Gollancz, 1963).

Tutt, Norman, 'A Decade of Policy', *British Journal of Criminology*, 21 (1981): 246–56.

US Department of Justice, *The Challenge of Crime in a Free Society: A Report by the President's Commission on Law Enforcement and Administration of Justice* (Washington DC, US Government Printing Office, 1967).

Vinson, Tony, *Wilful Obstruction: The Frustration of Prison Reform* (Melbourne, Methuen, 1982).

Wheeler, Stanton, Bonacish, Edna, Cramer, M. Richard, and Zola, Irving K., 'Agents of Delinquency Control: A Comparative Analysis', in Stanton Wheeler (ed.), *Controlling Delinquents* (New York, John Wiley, 1968), 31–60.

Williams, Raymond, *Culture and Society, 1780–1950* (London, Chatto and Windus, 1958).

Wilson, James Q., *Thinking About Crime* (New York, Vintage Books, 1975).

Woolf, Lord Justice, and Tumim, Judge Stephen, *Prison Disturbances April 1990*, Report of an Inquiry by Lord Justice Woolf and Judge Stephen Tumim, Cm. 1456 (London, HMSO, 1991).

Zimbardo, Philip G., Ebbeson, Ebbe B., and Maslach, Christine, *Influencing Attitudes and Changing Behaviour: An Introduction to Theory and Applications of Social Control and Personal Power* (New York, Random House, 1977).

Zimring, Franklin E., and Hawkins, Gordon, *The Scale of Imprisonment* (University of Chicago Press, 1991).

# INDEX

OXFORD

# MORE OXFORD PAPERBACKS

This book is just one of nearly 1000 Oxford Paperbacks currently in print. If you would like details of other Oxford Paperbacks, including titles in the World's Classics, Oxford Reference, Oxford Books, OPUS, Past Masters, Oxford Authors, and Oxford Shakespeare series, please write to:

**UK and Europe:** Oxford Paperbacks Publicity Manager, Arts and Reference Publicity Department, Oxford University Press, Walton Street, Oxford OX2 6DP.

Customers in UK and Europe will find Oxford Paperbacks available in all good bookshops. But in case of difficulty please send orders to the Cash-with-Order Department, Oxford University Press Distribution Services, Saxon Way West, Corby, Northants NN18 9ES. Tel: 0536 741519; Fax: 0536 746337. Please send a cheque for the total cost of the books, plus £1.75 postage and packing for orders under £20; £2.75 for orders over £20. Customers outside the UK should add 10% of the cost of the books for postage and packing.

**USA:** Oxford Paperbacks Marketing Manager, Oxford University Press, Inc., 200 Madison Avenue, New York, N.Y. 10016.

**Canada:** Trade Department, Oxford University Press, 70 Wynford Drive, Don Mills, Ontario M3C 1J9.

**Australia:** Trade Marketing Manager, Oxford University Press, G.P.O. Box 2784Y, Melbourne 3001, Victoria.

**South Africa:** Oxford University Press, P.O. Box 1141, Cape Town 8000.

# LAW FROM OXFORD PAPERBACKS

Oxford Paperbacks's law list ranges from introductions to the English legal system to reference books and in-depth studies of contemporary legal issues.

## INTRODUCTION TO ENGLISH LAW
### Tenth Edition

*William Geldart*
Edited by D. C. M. Yardley

'Geldart' has over the years established itself as a standard account of English law, expounding the body of modern law as set in its historical context. Regularly updated since its first publication, it remains indispensable to student and layman alike as a concise, reliable guide.

Since publication of the ninth edition in 1984 there have been important court decisions and a great deal of relevant new legislation. D. C. M. Yardley, Chairman of the Commission for Local Administration in England, has taken account of all these developments and the result has been a considerable rewriting of several parts of the book. These include the sections dealing with the contractual liability of minors, the abolition of the concept of illegitimacy, the liability of a trade union in tort for inducing a person to break his/her contract of employment, the new public order offences, and the intent necessary for a conviction of murder.

Other law titles:

*Freedom Under Thatcher: Civil Liberties in Modern Britain*
Keith Ewing and Conor Gearty
*Doing the Business*   Dick Hobbs
*Judges*   David Pannick
*Law and Modern Society*   P. S. Atiyah

# POLITICS IN OXFORD PAPERBACKS

Oxford Paperbacks offers incisive and provocative studies of the political ideologies and institutions that have shaped the modern world since 1945.

## GOD SAVE ULSTER!

### The Religion and Politics of Paisleyism

*Steve Bruce*

Ian Paisley is the only modern Western leader to have founded his own Church and political party, and his enduring popularity and success mirror the complicated issues which continue to plague Northern Ireland. This book is the first serious analysis of his religious and political careers and a unique insight into Unionist politics and religion in Northern Ireland today.

Since it was founded in 1951, the Free Presbyterian Church of Ulster has grown steadily; it now comprises some 14,000 members in fifty congregations in Ulster and ten branches overseas. The Democratic Unionist Party, formed in 1971, now speaks for about half of the Unionist voters in Northern Ireland, and the personal standing of the man who leads both these movements was confirmed in 1979 when Ian R. K. Paisley received more votes than any other member of the European Parliament. While not neglecting Paisley's 'charismatic' qualities, Steve Bruce argues that the key to his success has been his ability to embody and represent traditional evangelical Protestantism and traditional Ulster Unionism.

'original and profound . . . I cannot praise this book too highly.'
Bernard Crick, *New Society*

Also in Oxford Paperbacks:

*Freedom Under Thatcher*   Keith Ewing and Conor Gearty
*Strong Leadership*   Graham Little
*The Thatcher Effect*   Dennis Kavanagh and Anthony Seldon

## PHILOSOPHY IN OXFORD PAPERBACKS

Ranging from authoritative introductions in the Past Masters and OPUS series to in-depth studies of classical and modern thought, the Oxford Paperbacks' philosophy list is one of the most provocative and challenging available.

## THE GREAT PHILOSOPHERS

*Bryan Magee*

Beginning with the death of Socrates in 399, and following the story through the centuries to recent figures such as Bertrand Russell and Wittgenstein, Bryan Magee and fifteen contemporary writers and philosophers provide an accessible and exciting introduction to Western philosophy and its greatest thinkers.

Bryan Magee in conversation with:

| | |
|---|---|
| A. J. Ayer | John Passmore |
| Michael Ayers | Anthony Quinton |
| Miles Burnyeat | John Searle |
| Frederick Copleston | Peter Singer |
| Hubert Dreyfus | J. P. Stern |
| Anthony Kenny | Geoffrey Warnock |
| Sidney Morgenbesser | Bernard Williams |
| Martha Nussbaum | |

'Magee is to be congratulated . . . anyone who sees the programmes or reads the book will be left in no danger of believing philosophical thinking is unpractical and uninteresting.' Ronald Hayman, *Times Educational Supplement*

'one of the liveliest, fast-paced introductions to philosophy, ancient and modern that one could wish for' *Universe*

Also by Bryan Magee in Oxford Paperbacks:

*Men of Ideas*
*Aspects of Wagner* 2/e

## RELIGION AND THEOLOGY
## IN OXFORD PAPERBACKS

Oxford Paperbacks offers incisive studies of the philosophies and ceremonies of the world's major religions, including Christianity, Judaism, Islam, Buddhism, and Hinduism.

## A HISTORY OF HERESY

### *David Christie-Murray*

'Heresy, a cynic might say, is the opinion held by a minority of men which the majority declares unacceptable and is strong enough to punish.'

What is heresy? Who were the great heretics and what did they believe? Why might those originally condemned as heretics come to be regarded as martyrs and cherished as saints?

Heretics, those who dissent from orthodox Christian belief, have existed at all times since the Christian Church was founded and the first Christians became themselves heretics within Judaism. From earliest times too, politics, orthodoxy, and heresy have been inextricably entwined—to be a heretic was often to be a traitor and punishable by death at the stake—and heresy deserves to be placed against the background of political and social developments which shaped it.

This book is a vivid combination of narrative and comment which succeeds in both re-creating historical events and elucidating the most important—and most disputed—doctrines and philosophies.

Also in Oxford Paperbacks:

*Christianity in the West 1400–1700*   John Bossy
*John Henry Newman: A Biography*   Ian Ker
*Islam: The Straight Path*   John L. Esposito

# HISTORY IN OXFORD PAPERBACKS

Oxford Paperbacks' superb history list offers books on a wide range of topics from ancient to modern times, whether general period studies or assessments of particular events, movements, or personalities.

## THE STRUGGLE FOR
## THE MASTERY OF EUROPE 1848–1918

### A. J. P. Taylor

The fall of Metternich in the revolutions of 1848 heralded an era of unprecedented nationalism in Europe, culminating in the collapse of the Hapsburg, Romanov, and Hohenzollern dynasties at the end of the First World War. In the intervening seventy years the boundaries of Europe changed dramatically from those established at Vienna in 1815. Cavour championed the cause of *Risorgimento* in Italy; Bismarck's three wars brought about the unification of Germany; Serbia and Bulgaria gained their independence courtesy of the decline of Turkey—'the sick man of Europe'; while the great powers scrambled for places in the sun in Africa. However, with America's entry into the war and President Wilson's adherence to idealistic internationalist principles, Europe ceased to be the centre of the world, although its problems, still primarily revolving around nationalist aspirations, were to smash the Treaty of Versailles and plunge the world into war once more.

A. J. P. Taylor has drawn the material for his account of this turbulent period from the many volumes of diplomatic documents which have been published in the five major European languages. By using vivid language and forceful characterization, he has produced a book that is as much a work of literature as a contribution to scientific history.

'One of the glories of twentieth-century writing.' *Observer*

Also in Oxford Paperbacks:

*Portrait of an Age: Victorian England*   G. M. Young
*Germany 1866–1945*   Gorden A. Craig
*The Russian Revolution 1917–1932*   Sheila Fitzpatrick
*France 1848–1945*   Theodore Zeldin

# OPUS

*General Editors: Walter Bodmer, Christopher Butler, Robert Evans, John Skorupski*

OPUS is a series of accessible introductions to a wide range of studies in the sciences and humanities.

## METROPOLIS

### *Emrys Jones*

Past civilizations have always expressed themselves in great cities, immense in size, wealth, and in their contribution to human progress. We are still enthralled by ancient cities like Babylon, Rome, and Constantinople. Today, giant cities abound, but some are pre-eminent. As always, they represent the greatest achievements of different cultures. But increasingly, they have also been drawn into a world economic system as communications have improved.

*Metropolis* explores the idea of a class of supercities in the past and in the present, and in the western and developing worlds. It analyses the characteristics they share as well as those that make them unique; the effect of technology on their form and function; and the problems that come with size—congestion, poverty and inequality, squalor—that are sobering contrasts to the inherent glamour and attraction of great cities throughout time.

Also available in OPUS:

*The Medieval Expansion of Europe*   J. R. S. Phillips
*Metaphysics: The Logical Approach*   José A. Benardete
*The Voice of the Past* 2/e   Paul Thompson
*Thinking About Peace and War*   Martin Ceadel

# WOMEN'S STUDIES FROM
## OXFORD PAPERBACKS

Ranging from the *A–Z of Women's Health* to *Wayward Women: A Guide to Women Travellers*, Oxford Paperbacks cover a wide variety of social, medical, historical, and literary topics of particular interest to women.

## DESTINED TO BE WIVES
### The Sisters of Beatrice Webb
#### *Barbara Caine*

Drawing on their letters and diaries, Barbara Caine's fascinating account of the lives of Beatrice Webb and her sisters, the Potters, presents a vivid picture of the extraordinary conflicts and tragedies taking place behind the respectable façade which has traditionally characterized Victorian and Edwardian family life.

The tensions and pressures of family life, particularly for women; the suicide of one sister; the death of another, probably as a result of taking cocaine after a family breakdown; the shock felt by the older sisters at the promiscuity of their younger sister after the death of her husband are all vividly recounted. In all the crises they faced, the sisters formed the main network of support for each other, recognizing that the 'sisterhood' provided the only security in a society which made women subordinate to men, socially, legally, and economically.

Other women's studies titles:

*A–Z of Women's Health*  Derek Llewellyn-Jones
*'Victorian Sex Goddess': Lady Colin Campbell and the Sensational Divorce Case of 1886*  G. H. Fleming
*Wayward Women: A Guide to Women Travellers*
Jane Robinson
*Catherine the Great: Life and Legend*  John T. Alexander

# PAST MASTERS

## General Editor: Keith Thomas

The *Past Masters* series offers students and general readers alike concise introductions to the lives and works of the world's greatest literary figures, composers, philosophers, religious leaders, scientists, and social and political thinkers.

'Put end to end, this series will constitute a noble encyclopaedia of the history of ideas.' Mary Warnock

# HOBBES

## Richard Tuck

Thomas Hobbes (1588–1679) was the first great English political philosopher, and his book *Leviathan* was one of the first truly modern works of philosophy. He has long had the reputation of being a pessimistic atheist, who saw human nature as inevitably evil, and who proposed a totalitarian state to subdue human failings. In this new study, Richard Tuck shows that while Hobbes may indeed have been an atheist, he was far from pessimistic about human nature, nor did he advocate totalitarianism. By locating him against the context of his age, Dr Tuck reveals Hobbs to have been passionately concerned with the refutation of scepticism in both science and ethics, and to have developed a theory of knowledge which rivalled that of Descartes in its importance for the formation of modern philosophy.

Also available in Past Masters:

*Spinoza*   Roger Scruton
*Bach*   Denis Arnold
*Machiavelli*   Quentin Skinner
*Darwin*   Jonathan Howard

# OXFORD LETTERS AND MEMOIRS

Letters, memoirs, and journals offer a special insight into the private lives of public figures and vividly recreate the times in which they lived. This popular series makes available the best and most entertaining of these documents, bringing the past to life in a fresh and personal way.

## RICHARD HOGGART

### A Local Habitation
Life and Times: 1918–1940

With characteristic candour and compassion, Richard Hoggart evokes the Leeds of his boyhood, where as an orphan, he grew up with his grandmother, two aunts, an uncle, and a cousin in a small terraced back-to-back.

'brilliant . . . a joy as well as an education' Roy Hattersley

'a model of scrupulous autobiography' Edward Blishen, *Listener*

### A Sort of Clowning
Life and Times: 1940–1950

Opening with his wartime exploits in North Africa and Italy, this sequel to *A Local Habitation* recalls his teaching career in North-East England, and charts his rise in the literary world following the publication of *The Uses of Literacy*.

'one of the classic autobiographies of our time' Anthony Howard, *Independent on Sunday*

'Hoggart [is] the ideal autobiographer' Beryl Bainbridge, *New Statesman and Society*

Also in Oxford Letters and Memoirs:

*My Sister and Myself: The Diaries of J. R. Ackerley*
*The Letters of T. E. Lawrence*
*A London Family 1870–1900*   Molly Hughes